WHAT'S THE USE?

Also by Ian Stewart

Does God Play Dice?

Fearful Symmetry

Nature's Numbers

Life's Other Secret

Flatterland

The Annotated Flatland

Letters to a Young Mathematician

Why Beauty is Truth

Professor Stewart's Cabinet of Mathematical Curiosities

Mathematics of Life

Professor Stewart's Hoard of Mathematical Treasures

Seventeen Equations that Changed the World
(Alternative Title: *In Pursuit of the Unknown*)

The Great Mathematical Problems
(Alternative Title: *Visions of Infinity*)

Symmetry: A Very Short Introduction

Professor Stewart's Casebook of Mathematical Mysteries

The Beauty of Numbers in Nature

Professor Stewart's Incredible Numbers

Calculating the Cosmos

Significant Figures

Do Dice Play God?

WHAT'S THE USE?

The Unreasonable Effectiveness of Mathematics

IAN STEWART

P

PROFILE BOOKS

First published in Great Britain in 2021 by
Profile Books Ltd
29 Cloth Fair
London
EC1A 7JQ

www.profilebooks.com

Copyright © Joat Enterprises, 2021

1 3 5 7 9 10 8 6 4 2

The moral right of the author has been asserted.

All rights reserved. Without limiting the rights under copyright
reserved above, no part of this publication may be reproduced,
stored or introduced into a retrieval system, or transmitted, in
any form or by any means (electronic, mechanical, photocopying,
recording or otherwise), without the prior written permission of
both the copyright owner and the publisher of this book.

All reasonable efforts have been made to obtain copyright permissions where
required. Any omissions and errors of attribution are unintentional and
will, if notified in writing to the publisher, be corrected in future printings.

A CIP catalogue record for this book is available from the British Library.

ISBN 978 1 78125 941 2
eISBN 978 1 78283 400 7
Export ISBN 978 1 78816 807 6

Typeset in Sabon by MacGuru Ltd
Printed and bound in Britain by Clays Ltd, Elcograf S.p.A.

FSC
www.fsc.org
MIX
Paper from
responsible sources
FSC® C018072

Contents

1

Unreasonable Effectiveness

> The miracle of the appropriateness of the language of
> mathematics for the formulation of the laws of physics
> is a wonderful gift which we neither understand nor
> deserve. We should be grateful for it and hope that it will
> remain valid in future research and that it will extend, for
> better or for worse, to our pleasure even though perhaps
> also to our bafflement, to wide branches of learning.
>
> Eugene Wigner, *The Unreasonable Effectiveness of
> Mathematics in the Natural Sciences*

What is mathematics for?

What is it doing for *us*, in our daily lives?

Not so long ago, there were easy answers to these questions. The typical citizen used basic arithmetic all the time, if only to check the bill when shopping. Carpenters needed to know elementary geometry. Surveyors and navigators needed trigonometry as well. Engineering required expertise in calculus.

Today, things are different. The supermarket checkout totals the bill, sorts out the special meal deal, adds the sales tax. We listen to the beeps as the laser scans the barcodes, and as long as the beeps match the goods, we assume the electronic gizmos know what they're doing. Many professions still rely on extensive mathematical knowledge, but even there, we've outsourced most of the mathematics to electronic devices with built-in algorithms.

My subject is conspicuous by its absence. The elephant isn't even in the room.

It would be easy to conclude that mathematics has become outdated and obsolete, but that view is mistaken. Without mathematics, today's world would fall apart. As evidence, I'm going to show you applications to politics, the law, kidney transplants, supermarket delivery schedules, Internet security, movie special effects, and making springs. We'll see how mathematics plays an essential role in medical scanners, digital photography, fibre broadband, and satellite navigation. How it helps us predict the effects of climate change; how it can protect us against terrorists and Internet hackers.

Remarkably, many of these applications rely on mathematics that originated for totally different reasons, often just the sheer fascination of following your nose. While researching this book I was repeatedly surprised when I came across uses of my subject that I'd never dreamed existed. Often they exploited topics that I wouldn't have expected to have practical applications, like space-filling curves, quaternions, and topology.

Mathematics is a boundless, hugely creative system of ideas and methods. It lies just beneath the surface of the transformative technologies that are making the twenty-first century totally different from any previous era – video games, international air travel, satellite communications, computers, the Internet, mobile phones.[1] Scratch an iPhone, and you'll see the bright glint of mathematics.

Please don't take that literally.

*

There's a tendency to assume that computers, with their almost miraculous abilities, are making mathematicians, indeed mathematics itself, obsolete. But computers no more displace mathematicians than the microscope displaced biologists. Computers change the way we go about *doing* mathematics, but mostly they relieve us of the tedious bits. They give us time to think, they help us search for patterns, and they add a powerful new weapon to help advance the subject more rapidly and more effectively.

In fact, a major reason why mathematics is becoming ever more essential is the ubiquity of cheap, powerful computers. Their rise has opened up new opportunities to apply mathematics to real-world issues. Methods that were hitherto impractical, because they needed too many calculations, have now become routine. The greatest mathematicians of the pencil-and-paper era would have flung up their hands in despair at any method requiring a billion calculations. Today, we routinely use such methods, because we have technology that can do the sums in a split second.

Mathematicians have long been at the forefront of the computer revolution – along with countless other professions, I hasten to add. Think of George Boole, who pioneered the symbolic logic that forms the basis of current computer architecture. Think of Alan Turing, and his universal Turing machine, a mathematical system that can compute anything that's computable. Think of Muhammad al-Khwarizmi, whose algebra text of AD 820 emphasised the role of systematic computational procedures, now named after him: *algorithms*.

Most of the algorithms that give computers their impressive abilities are firmly based on mathematics. Many of the techniques concerned have been taken 'off the shelf' from the existing store of mathematical ideas, such as Google's PageRank algorithm, which quantifies how important a website is and founded a multi-billion dollar industry. Even the snazziest deep learning algorithm in artificial intelligence uses tried and tested mathematical concepts such as matrices and weighted graphs. A task as prosaic as searching a document for a particular string of letters involves, in one common method at least, a mathematical gadget called a finite-state automaton.

The involvement of mathematics in these exciting developments tends to get lost. So next time the media propel some miraculous new ability of computers to centre stage, bear in mind that hiding in the wings there will be a lot of mathematics, *and* a lot of engineering, physics, chemistry, and psychology as well, and that without the support of this hidden cast of helpers, the digital superstar would be unable to strut its stuff in the spotlight.

*

The importance of mathematics in today's world is easily underestimated because nearly all of it goes on behind the scenes. Walk along a city street and you're overwhelmed by signs proclaiming the daily importance of banks, greengrocers, supermarkets, fashion outlets, car repairs, lawyers, fast food, antiques, charities, and a thousand other activities and professions. You don't find a brass plaque announcing the presence of a consulting mathematician. Supermarkets don't sell you mathematics in a can.

Dig a little deeper, however, and the importance of mathematics quickly becomes apparent. The mathematical equations of aerodynamics are vital to aircraft design. Navigation depends on trigonometry. The way we use it today is different from how Christopher Columbus used it, because we embody the mathematics in electronic devices instead of pen, ink, and navigation tables, but the underlying principles are much the same. The development of new medicines relies on statistics to make sure the drugs are safe and effective. Satellite communications depend on a deep understanding of orbital dynamics. Weather forecasting requires the solution of equations for how the atmosphere moves, how much moisture it contains, how warm or cold it is, and how all of those features interact. There are thousands of other examples. We don't notice they involve mathematics, because we don't need to know that to benefit from the results.

What makes mathematics so useful, in such a broad variety of human activities?

It's not a new question. In 1959 the physicist Eugene Wigner gave a prestigious lecture at New York University,[2] with the title 'The Unreasonable Effectiveness of Mathematics in the Natural Sciences'. He focused on science, but the same case could have been made for the unreasonable effectiveness of mathematics in agriculture, medicine, politics, sport ... you name it. Wigner himself hoped that this effectiveness would extend to 'wide branches of learning'. It certainly did.

The key word in his title stands out because it's a surprise: *unreasonable*. Most uses of mathematics are entirely reasonable, once you find out which methods are involved in solving an important problem or inventing a useful gadget. It's entirely reasonable, for instance, that engineers use the equations of aerodynamics to help them design aircraft. That's what aerodynamics was created for in the first place. Much of the mathematics used in weather forecasting arose with that purpose in mind. Statistics emerged from the discovery of large-scale patterns in data about human behaviour. The amount of mathematics required to design spectacles with varifocal lenses is huge, but most of it was developed with optics in mind.

The ability of mathematics to solve important problems becomes unreasonable, in Wigner's sense, when no such connection exists between the original motivation for developing the mathematics, and the eventual application. Wigner began his lecture with a story, which I'll paraphrase and embellish slightly.

Two former school classmates met up. One, a statistician working on population trends, showed the other one of his research papers, which began with a standard formula in statistics, the normal distribution or 'bell curve'.[3] He explained various symbols – this one is the population size, that one is a sample average – and how the formula can be used to infer the size of the population without having to count everyone. His classmate suspected his friend was joking, but he wasn't entirely sure, so he asked about other symbols. Eventually he came to one that looked like this: π.

'What's that? It looks familiar.'

5

'Yes, it's *pi* – the ratio of the circumference of the circle to its diameter.'

'Now I know you're pulling my leg,' said the friend. 'What on earth can a circle have to do with population sizes?'

The first point about this story is that the friend's scepticism was entirely sensible. Common sense tells us that two such disparate concepts can't possibly be related. One is about geometry, the other about people, for heaven's sake. The second point is that despite common sense, there's a connection. The bell curve has a formula, which happens to involve the number π. It's not just a convenient approximation; the exact number really is good old familiar π. But the reason it appears in the context of the bell curve is far from intuitive, even to mathematicians, and you need advanced calculus to see how it arises, let alone *why*.

Let me tell you another story about π. Some years ago we had the downstairs bathroom renovated. Spencer, an amazingly versatile craftsman who came to fit the tiles, discovered that I wrote popular mathematics books. 'I've got a maths problem for you,' he said. 'I've got to tile a circular floor, and I need to know its area to work out how many tiles I'll need. There was some formula they taught us …'

'Pi *r* squared,' I replied.

'That's the one!' So I reminded him how to use it. He went away happy, with the answer to his tiling problem, a signed copy of one of my books, and the discovery that the mathematics he'd done at school was, contrary to his long-held belief, useful in his present occupation.

The difference between the two stories is clear. In the second story, π turns up because it was introduced to solve exactly that kind of problem in the first place. It's a simple, direct story about the effectiveness of mathematics. In the first story, π also turns up and solves the problem, but its presence is a surprise. It's a story of *unreasonable* effectiveness: an application of a mathematical idea to an area totally divorced from that idea's origins.

*

In *What's the Use?* I'm not going to say much about reasonable uses of my subject. They're worthy, they're interesting, they're as much a part of the mathematical landscape as everything else, they're equally important – but they don't make us sit up and say 'Wow!' They can also mislead the Powers That Be into imagining that the only way to advance mathematics is to decide on the problems and then get the mathematicians to invent ways to solve them. There's nothing wrong with goal-oriented research of this kind, but it's fighting with one arm tied behind your back. History repeatedly shows the value of the second arm, the amazing scope of human imagination. What gives mathematics its power is the *combination* of these two ways of thinking. Each complements the other.

For instance, in 1736, the great mathematician Leonhard Euler turned his mind to a curious little puzzle about people taking walks across bridges. He knew it was interesting, because it seemed to require a new kind of geometry, one that abandoned the usual ideas of lengths and angles. But he couldn't possibly have anticipated that in the twenty-first century the subject that his solution kick-started would help more patients get life-saving kidney transplants. For a start, kidney transplants would have seemed pure fantasy at that time, but even if they hadn't, any connection with the puzzle would have seemed ridiculous.

And who would ever have imagined that the discovery of space-filling curves – curves that pass through *every* point of a solid square – could help Meals on Wheels to plan its delivery routes? Certainly not the mathematicians who studied such questions in the 1890s, who were interested in how to define esoteric concepts like 'continuity' and 'dimension', and initially found themselves explaining why cherished mathematical beliefs can be wrong. Many of their colleagues denounced the entire enterprise as wrong-headed and negative. Eventually everyone realised that it's no good living in a fool's paradise, assuming that everything will work perfectly when in fact it won't.

It's not just the mathematics of the past that gets used in this way. The kidney transplant methods rely on numerous modern extensions of Euler's original insight, among them powerful algorithms for combinatorial optimisation – making the best choice from a huge range of possibilities. The myriad mathematical techniques employed by movie animators include many that go back a decade or less. An example is 'shape space', an infinite-dimensional space of curves that are considered to be the same if they differ only by a change of coordinates. They're used to make animation sequences appear smoother and more natural. Persistent homology, another very recent development, arose because pure mathematicians wanted to compute complicated topological invariants that count multidimensional holes in geometric shapes. Their method also turned out to be an effective way to ensure that sensor networks provide full coverage when protecting buildings or military bases against terrorists or other criminals. Abstract concepts from algebraic geometry – 'supersingular isogeny graphs' – can make Internet communications secure against quantum computers. These are so new that they currently exist only in rudimentary form, but they will trash today's cryptosystems if they can fulfil their potential.

Mathematics doesn't just spring such surprises on rare occasions. It makes a positive habit of it. In fact, as far as many mathematicians are concerned, these surprises are the most interesting uses of the subject, and the main justification for considering it to *be* a subject, rather than just a rag-bag of assorted tricks, one for each kind of problem.

Wigner went on to say that 'the enormous usefulness of mathematics in the natural sciences is something bordering on the mysterious and … there is no rational explanation for it.' It is, of course, true that mathematics grew out of problems in science in the first place, but Wigner wasn't puzzled by the subject's effectiveness in areas it was designed for. What baffled him was its effectiveness in apparently unrelated ones. Calculus grew from Isaac Newton's research on the motion of the planets, so it's not greatly surprising that it helps us to understand how planets move. However, it *is* surprising when calculus lets us make statistical estimates of human populations, as in Wigner's little story, explains changes in the numbers of fish caught in the Adriatic Sea during the First World War,[4] governs option pricing in the financial sector, helps engineers to design passenger jets, and is vital for telecommunications. Because calculus wasn't invented for any such purpose.

Wigner was right. The way mathematics repeatedly turns up uninvited in the physical sciences, and in most other areas of human activity, is a mystery. One proposed solution is that the universe is 'made of' mathematics, and humans are just digging out this basic ingredient. I'm not going to argue the toss here, but if this explanation is correct it replaces one mystery by an even deeper one. *Why* is the universe made of mathematics?

*

On a more pragmatic level, it can be argued that mathematics has several features that help to make it unreasonably effective in Wigner's sense. One is, I agree, its many links with natural science, which transfer to the human world as transformative technology. Many of the great mathematical innovations have indeed arisen from scientific enquiries. Others are rooted in human concerns. Numbers arose from basic accountancy (how many sheep have I got?). Geometry *means* 'earth-measurement', and was closely related to the taxation of land and in ancient Egypt to the construction of pyramids. Trigonometry emerged from astronomy, navigation, and map-making.

However, that alone isn't an adequate explanation. Many other great mathematical innovations have *not* arisen from scientific enquiry or specific human issues. Prime numbers, complex numbers, abstract algebra, topology – the primary motivation for these discoveries/inventions was human curiosity and a sense of pattern. This is a second reason why mathematics is so effective: mathematicians use it to seek patterns and to tease out underlying structure. They search for *beauty*, not of form but of logic. When Newton wanted to understand the motion of the planets, the solution came when he thought like a mathematician and looked for deeper patterns beneath the raw astronomical data. Then he came up with his law of gravity.[5] Many of the greatest mathematical ideas had no real-world motivation at all. Pierre de Fermat, a lawyer who did mathematics for fun in the seventeenth century, made fundamental discoveries in number theory: deep patterns in the behaviour of ordinary whole numbers. It took three centuries for his work in that area to acquire practical applications, but right now the commercial transactions that drive the Internet wouldn't be possible without it.

Another feature of mathematics that's become increasingly evident since the late 1800s is *generality*. Different mathematical structures have many common features. The rules of elementary algebra are the same as those of arithmetic. Different kinds of geometry (Euclidean, projective, non-Euclidean, even topology) are all closely related to each other. This hidden unity can be made explicit by working, from the start, with general structures that obey specified rules. Understand the generalities, and all the special examples become obvious. This saves a lot of effort, which would otherwise be wasted doing essentially the same thing many times in slightly different language. It has one downside, however: it tends to make the subject more abstract. Instead of talking about familiar things such as numbers, the generalities must refer to anything obeying the same *rules* as numbers, with names like 'Noetherian ring', 'tensor category', or 'topological vector space'. When this kind of abstraction is carried to extremes, it can become difficult to understand what the generalities *are*, let alone how to make use of them. Yet they're so useful that the human world would no longer function without them. You want Netflix? Someone has to do the maths. It's not magic; it just feels like it.

A fourth feature of mathematics, highly relevant to this discussion, is its *portability*. This is a consequence of its generality, and it's why abstraction is necessary. Irrespective of the problem that motivated it, a mathematical concept or method possesses a level of generality that often makes it applicable to quite different problems. Any problem that can be recast in the appropriate framework then becomes fair game. The simplest and most effective way to create portable mathematics is to design portability in from the start, by making the generalities explicit.

For the last two thousand years, mathematics has taken its inspiration from three main sources: the workings of nature, the workings of humanity, and the internal pattern-seeking tendencies of the human mind. These three pillars support the entire subject. The miracle is that despite its multifarious motivations, mathematics is *all one thing*. Every branch of the subject, whatever its origins and aims, has become tightly bound to every other branch – and the bonds are becoming ever stronger and ever more entangled.

This points to a fifth reason why mathematics is so effective, and in such unexpected ways: its *unity*. And alongside this goes a sixth, for which I'll provide ample evidence as we proceed: its *diversity*.

Reality, beauty, generality, portability, unity, diversity. Which, together, lead to utility.

It's as simple as that.

2

How Politicians Pick Their Voters

Ankh-Morpork had dallied with many forms of
government and had ended up with that form of
democracy known as One Man, One Vote. The Patrician
was the Man; he had the Vote.

Terry Pratchett, *Mort*

The ancient Greeks gave the world many things – poetry, drama,
sculpture, philosophy, logic. They also gave us geometry and
democracy, which have turned out to be more closely linked than
anyone might have expected, least of all the Greeks. To be sure,
the political system of ancient Athens was a very limited form
of democracy; only free men could vote, not women or slaves.
Even so, in an era dominated by hereditary rulers, dictators, and
tyrants, Athenian democracy was a distinct advance. As was Greek
geometry, which, in the hands of Euclid of Alexandria, emphasised the importance of making your basic assumptions clear and
precise, and deriving everything else from them in a logical and
systematic fashion.

How on earth can mathematics be applied to politics? Politics is about human relationships, agreements, and obligations,
whereas mathematics is about cold, abstract logic. In political circles, rhetoric trumps logic, and the inhuman calculations
of mathematics seem far removed from political bickering. But
democratic politics is carried out according to rules, and rules
have consequences that aren't always foreseen when they're first

drawn up. Euclid's pioneering work in geometry, collected in his famous *Elements*, set a standard for deducing consequences from rules. In fact, that's not a bad definition of mathematics as a whole. At any rate, after a mere 2,500 years, mathematics is beginning to infiltrate the political world.

One of the curious features of democracy is that politicians who claim to be devoted to the idea that decisions should be made by 'The People' repeatedly go out of their way to ensure that this doesn't happen. This tendency goes right back to the first democracy in ancient Greece, where the right to vote was given only to adult male Athenians, about one third of the adult population. From the moment the idea of electing leaders and selecting policy by popular vote was conceived, so was the even more attractive idea of subverting the entire process, by controlling who voted and how effective their votes are. This is easy, even when every voter gets one vote, because the effectiveness of a vote depends on the context in which it's cast, and you can rig the context. As journalism professor Wayne Dawkins delicately put it, this amounts to politicians picking their voters instead of voters picking their politicians.[6]

That's where mathematics comes in. Not in the cut-and-thrust of political debate, but in the structure of the debating rules and the context in which they apply. Mathematical analysis cuts both ways. It can reveal new, cunning methods for rigging votes. It can also shine a spotlight on such practices, providing clear evidence of that kind of subversion, which can sometimes be used to prevent it happening.

Mathematics also tells us that any democratic system must involve elements of compromise. You can't have everything you want, however desirable that might be, because the list of desirable attributes is self-contradictory.

*

On 26 March 1812 the *Boston Gazette* gave the world a new word: *gerrymander*. Originally spelt Gerry-mander, it's what Lewis Carroll later called a portmanteau word, created by combining two standard words. 'Mander' was the tail end of 'salamander', and 'Gerry' was the tail end of Elbridge Gerry, governor of Massachusetts. We don't know for sure who first put the two tails together, but on circumstantial grounds, historians tend to opt for one of the newspaper's editors, Nathan Hale, Benjamin Russell, or John Russell. Incidentally, 'Gerry' was pronounced with a hard G, like 'Gary', but 'gerrymander' has a soft G, like 'Jerry'.

What was Elbridge Gerry doing that got him combined with a lizard-like creature, which, in medieval folklore, was reputed to dwell in fire?

Rigging an election.

More precisely, Gerry was responsible for a bill that redrew the district boundaries in Massachusetts for elections to the state senate. Districting, as it's called, naturally leads to boundaries; it is, and has long been, common in most democracies. The overt reason is practicality: it's awkward to take decisions if the entire nation gets to vote on every proposal. (Switzerland comes close: up to four times a year the Federal Council chooses proposals for citizens to vote on, essentially a series of referendums. On the other hand, women didn't get the vote there until 1971, and one canton held out until 1991.) The time-honoured solution is for voters to elect a much smaller number of representatives, and let the representatives make the decisions. One of the fairer methods is proportional representation: the number of representatives of a given political party is proportional to the number of votes that that party receives. More commonly, the population is divided into districts, and each district elects a number of representatives, roughly proportional to the number of electors in that district.

For example, in American presidential elections, each state votes for a specific number of 'electors' – members of the Electoral College. Each elector has one vote, and who becomes President is decided by a simple majority of these votes. It's a system that originated when the only way to get a message from the American hinterland to the centres of power was to carry a letter on horseback or in a horse-drawn coach. Long-distance rail and the telegraph came later. In those days, totalling up the votes of huge numbers of individuals was too slow.[7] But this system also ceded control to the elite members of the Electoral College. In British parliamentary elections, the country is divided into (mainly geographical) constituencies, each of which elects one Member of Parliament (MP). Then the party (or combination of parties in a coalition) with the most MPs forms the government, and chooses one of its MPs to be Prime Minister, by a variety of methods. The Prime Minister has considerable powers, and in many ways is more like a President.

There's also a covert reason for funnelling democratic decisions through a small number of gatekeepers: it's easier to rig the vote. All such systems have innate flaws, which often lead to strange results, and on occasion they can be exploited to flout the Will of the People. In several recent US presidential elections, the total number of votes cast by the People for the candidate who lost was greater than the number of votes for the candidate who won. Agreed, the current method for choosing a President doesn't depend on the popular vote, but with modern communications the only reason not to change to a fairer system is that a lot of powerful people prefer it the way it is.

The underlying problem here is 'wasted votes'. In each state, a candidate needs half the total plus one vote (or half a vote if the total is odd) to win; any extra votes beyond that threshold make no difference to what happens at the Electoral College stage. Thus, in the 2016 presidential election, Donald Trump received 304 votes in the Electoral College compared to 227 for Hillary Clinton, but Clinton's popular vote exceeded Trump's by 2·87 million. Trump thereby became the fifth US President to be elected while losing the popular vote.

The Gerry-mander, thought to have been drawn in 1812 by Elkanah Tisdale.

The boundaries of American states are effectively immutable, so this is not a districting issue. In other elections, the boundaries of districts can be redrawn, usually by the party in power, and a more insidious flaw appears. Namely, that party can draw the boundaries to ensure that unusually large numbers of votes for the opposing party are wasted. Cue Elbridge Gerry and the senate vote. When Massachusetts voters saw the map of electoral districts, most of them looked entirely normal. One didn't. It combined twelve counties from the west and north of the state into a singe, sprawling, meandering region. To the political cartoonist responsible for the drawing that shortly appeared in the *Boston Gazette* – probably the painter, designer, and engraver Elkanah Tisdale – this district closely resembled a salamander.

Gerry belonged to the Democratic-Republican Party, which was in competition with the Federalists. In the 1812 election the Federalists won the House and governorship in the state, which put Gerry out of office. However, his redistricting of the state senate worked a treat, and it was held comfortably by the Democratic-Republicans.

*

The mathematics of gerrymandering begins by looking at how people do it. There are two main tactics, packing and cracking. *Packing* spreads your own vote as evenly as possible, with a small but decisive majority, in as many districts as possible, and cedes the rest to the enemy. Sorry, opposition. *Cracking* breaks up the opposition's votes so that they lose as many districts as possible. *Proportional representation*, in which the number of representatives is proportional to each party's total votes (or as close to that as possible given the numbers) avoids these tricks, and is fairer. Unsurprisingly, the US constitution makes proportional representation illegal, because as the law stands, districts must have only one representative. In 2011 the UK held a referendum on another alternative, the single transferable vote: the people voted against this change. There's never been a referendum on proportional representation in the UK.

Here's how packing and cracking work, in an artificial example with very simple geography and voting distributions.

The state of Jerimandia is contested by two political parties, the Lights and the Darks. There are fifty regions, to be split into five districts. In recent elections, Light has a majority in twenty

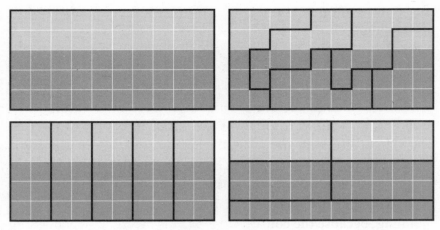

Carving up Jerimandia. *Top left*: Fifty regions to be cut into five districts of ten each. Voters are known to prefer the Light or Dark party according to the shading. *Top right*: Packing gives Light three districts and Dark only two. *Bottom left*: Cracking gives Dark all five districts. *Bottom right:* This arrangement would give proportional representation.

of them, all in the north, while Dark has a majority in the thirty southern regions (top left). The Light administration, which just scraped in at the previous vote, has redistricted the state by packing more of its voters into three of the districts (top right) so that it wins three and Dark gets only two. Dark subsequently challenges this redistricting in court, on the grounds that the shapes of the districts are obviously gerrymandered, and manages to gain control of redistricting for the next election, when it uses cracking (bottom left) to ensure that it will win all five districts.

If the districts must be composed of ten of the small square regions, the best that Light can achieve by packing is three regions out of the five. They need to win six out of ten regions to win a district, and they control twenty regions; that gives them three sixes plus a two, which is wasted. The best that Dark can achieve by cracking is all five. Proportional representation would give Light two districts and Dark three, like the bottom right picture.

(In practice, proportional representation is not achieved by districting.)

*

Countries ruled by dictators, or what amount to dictators, commonly run elections to prove to the world how democratic they are. These elections are generally rigged, and even if legal challenges are permitted, they never succeed because the courts are rigged too. In other countries, it's not only possible to challenge any particular instance of redistricting, but there's a chance of winning, because the court's judgement is mostly independent of the governing party. Except for appointments of judges on a partisan basis, of course.

In such cases the main problem facing the judges isn't political. It's to find objective ways to assess whether gerrymandering has occurred. For every 'expert' who eyeballs the map and declares a gerrymander, you can always find another who comes to the opposite conclusion. More objective methods than opinions and verbal arguments are needed.

This is a clear opportunity for mathematics. Formulas or algorithms can quantify whether district boundaries are reasonable and fair, or artificial and biased, in some clearly defined sense. The design of these formulas or algorithms is not of itself an objective process, of course, but once they're agreed upon (partly a political process) everyone concerned knows what they are, and their results can be verified independently. This provides the court with a logical basis for its decision.

Having understood the underhand methods that politicians can use to implement partisan redistricting, you can invent mathematical quantities or rules to detect them. No such rule can be perfect – in fact, there's a proof that this is impossible, which I'll come to once we have the background to appreciate what it tells us. There are five types of approach in current use:

- Detect strangely shaped districts.
- Detect imbalances in the proportion of seats to votes.
- Quantify how many wasted votes a given division creates, and compare that to what has been legally decided to be acceptable.
- Consider all possible electoral maps, estimate the likely outcome in terms of seats based on existing voter data, and see if the proposed map is a statistical outlier.
- Set up protocols that guarantee the eventual decision is fair, is seen to be fair, and is agreed to be fair by both parties.

The fifth approach is the most surprising, and the surprise is that it can actually be done. Let's take them in turn, saving the surprise till last.

*

First, weird shapes.

As long ago as 1787 James Madison wrote in *The Federalist Papers* that 'the natural limit of a democracy is that distance from the central point which will just permit the most remote citizens to assemble as often as their public functions demand'. Taken literally, he was proposing that districts should be roughly circular, and not so large that travel times from the periphery to the centre become unreasonable.

Suppose, for instance, that the main support for a political party is concentrated in coastal regions. Including all of these voters in a single district leads to a long, thin, winding shape, running all the way along the coast – completely unnatural compared to all the other nice, compact, sensible districts. It would be hard not to conclude that some funny business had been going on, and that the boundaries have been drawn to ensure that a lot of that party's votes are wasted. Gerrymandered districts often betray their partisan nature by their weird shapes, as did the original district that led to the name.

The legal profession can argue till the cows come home about what constitutes a weird shape. So in 1991 lawyers Daniel Polsby and Robert Popper proposed a way to quantify how weird a shape is, now known as its Polsby–Popper score.[8] This is calculated as

$$4\pi \text{ times area of district/square of perimeter of district}$$

Anyone with any mathematical sensitivity is immediately drawn to that factor 4π. Just as Wigner's friend wondered how populations can have anything to do with circles, we can ask what circles have to do with political districting. The answer is refreshingly simple and direct: the circle is the most compact possible region.

This fact has a lengthy history. According to ancient Greek and Roman sources, notably Virgil's epic poem the *Aeneid* and the *Philippic Histories* of Gnaeus Pompeius Trogus, the founder of the city state of Carthage was Queen Dido. Trogus's historical account was summarised by Junianus Justinus in the third century AD, and it tells a striking legend. Dido and her brother Pygmalion were the joint heirs of an unnamed king of the city of Tyre. When the king died, the people wanted Pygmalion to rule on his own, despite his youth. Dido married her uncle Acerbas, who was rumoured to have a secret hoard of wealth, which Pygmalion wanted, so he murdered Acerbas. Dido pretended to throw the supposed hoard of gold into the sea, although in reality she dumped bags of sand. Sensibly fearing Pygmalion's wrath, she fled, first to Cyprus and then to the north coast of Africa. She asked the Berber king Iarbas to grant her a small piece of land where she could rest for a time, and he agreed that she could have as much land as could be surrounded by an oxhide. Dido cut the hide into very thin strips, and made a circle round a nearby hill, which to this day is called Byrsa, meaning 'hide'. The settlement became the city of Carthage, and when it grew wealthy, Iarbas told Dido that she must marry him or face the city's destruction. She sacrificed many victims on a huge pyre, pretending this was to honour her first husband, to ready her for marriage to Iarbas; then she climbed the pyre, said that she would join her first husband rather than submit to Iarbas's desires, and killed herself with a sword.

We don't know whether Dido actually existed, though Pygmalion definitely did, and some sources mention Dido as well as him. It's therefore pointless to enquire about the historical accuracy of the legend. Be that as it may, the historical legend has a mathematical one concealed within it: Dido used the hide to *encircle* the hill. Why a circle? Because – so mathematicians claim – she knew that a circle encloses the greatest area for a given circumference.[9] This fact, which bears the impressive name 'isoperimetric inequality', was known empirically in ancient Greece, but not proved rigorously until 1879, when the complex analyst Karl Weierstrass filled a gap in five different proofs published by the geometer Jakob Steiner. Steiner proved that if an optimal shape exists, it must be the circle, but he failed to prove existence.[10]

The isoperimetric inequality states that

square of perimeter is greater than or equal to 4π times area

This applies to any shape in the plane that's sufficiently well behaved to have a perimeter and an area. Moreover, the constant 4π is best possible – it can't be made larger – and 'greater than or equal to' becomes equality only when the shape is a circle.[11] The isoperimetric inequality led Polsby and Popper to suggest that the quantity I named the Polsby–Popper (PP) score is an effective way to measure how round a shape is. For example, the score for the following shapes is:

Circle: PP score = 1
Square: PP score = 0·78
Equilateral triangle: PP score = 0·6

The Gerry-mander's PP score is about 0·25.

However, the PP score has serious flaws. Weird shapes can sometimes be unavoidable because of the local geography, such as rivers, lakes, forests, and the shapes of coastlines. Moreover, a district can be neat and compact, yet clearly gerrymandered. A 2011 boundary plan for voting for the Pennsylvania state legislature was very distorted and artificial, so in 2018 Republicans in the state legislature drew up proposals to replace it. The draft districts were rated strongly compact according to five measures that the state's Supreme Court had specified, but a mathematical analysis of voter distributions within those regions showed that the boundaries were highly partisan and would bias the voting results.

Even the scale on which a map is drawn can cause problems. The main issue here is fractal geometry. A fractal is a geometric shape with detailed structure on all scales. Many natural shapes look like fractals – at least, they look a lot more like them than they do Euclid's triangles and circles. Coastlines and clouds can be modelled very effectively as fractals, which capture their intricate structure. The word was coined in 1975 by Benoit Mandelbrot, who pioneered and promoted the whole area of fractal geometry. Coastlines and rivers are extremely wiggly fractal curves, and the length you measure depends on how fine a scale you use to make the measurement. Indeed, the length of a fractal curve is technically infinite, which translates into everyday reality as 'the measured length grows without limit as you look ever more closely'. So lawyers can argue indefinitely about the measurement of the perimeter, let alone whether the district was gerrymandered.

*

Since the weirdness of a shape is so tricky, let's try something more straightforward. Do the results of votes match the statistical voting patterns of the electorate?

What's the Use?

If there are ten seats up for grabs and voters split 60–40, you might expect 6 seats for one party and 4 for the other. If one party wins the lot, you might suspect gerrymandering. But it's not that simple. This type of outcome is common in 'first past the post' voting systems. In the UK's general election of 2019, the Conservative party received 44% of the vote but won 365 seats out of 650, which is 56% of the seats. Labour got 32% of the vote and 31% of the seats. The Scottish Nationalists, with 4% of the vote, got 7% of the seats (though this is a special case since its voter base is entirely in Scotland). The Liberal Democrats had 12% of the vote and 2% of the seats. Most of these discrepancies resulted from regional voting patterns, not from strangely drawn boundaries. After all, if a two-party election for one person, say a President, is decided by a simple majority, 50% of the votes (plus one) will secure 100% of the office.

Here's an American example. In Massachusetts, in federal and presidential elections since 2000, the Republicans have secured more than a third of votes overall. Yet the last time a Republican won a seat on the House of Representatives in that state was 1994. Gerrymandering? Probably not. If the one third of Republican voters are distributed fairly uniformly across the state, then however you draw the district boundaries – except for ridiculous shapes winding around and between the homes of individual citizens – the proportion of Republican voters in any district will remain roughly one third. Democrats will win the lot. And that's exactly what happened.

In one real-world election, mathematicians have shown that this kind of effect can be unavoidable, however the boundaries are drawn – at least, without splitting up individual towns. In 2006 Kenneth Chase challenged Edward Kennedy for the US Senate, when Massachusetts was divided into nine congressional districts. Chase got 30% of the total vote, but lost in all nine districts. Computer analysis of the possibilities showed that no single district-sized collection of towns, even if scattered irregularly

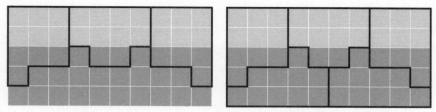

Left: Light's proposal, with two districts left for Dark to define.
Right: The most compact choice they could make.

across the state, would have given Chase a win. Chase's supporters were distributed fairly uniformly in most towns; you couldn't
have gerrymandered him a win *whatever* boundary you drew.

Back in Jerimandia, when Dark won all five districts, Light
objected to this particular redistricting on the grounds that the
rectangular districts were too long and thin, so Dark had obviously indulged in cracking. The court ruled that the districts
should be more compact. Light proposed three compact districts,
and generously offered to let Dark choose how to divide what was
left into two more districts. Dark objected because this gave Light
three districts and Dark only two, despite Dark having more of
the vote.

This division reveals two further flaws in the use of compactness to detect gerrymandering. Although it arguably *is* compact
– so far – it gives Light three fifths of the districts with two fifths
of the vote. Moreover, there's no way to divide what's left into two
compact districts. The geography of Jerimandia makes it difficult
to achieve compactness and fairness at the same time. Perhaps
impossible, depending on the definitions.

*

Since compactness is flawed, what else can we do to spot partisan redistricting? Voting data tell us not just the outcome of an

election, but what it would have been if the votes cast for each party were shifted by specific amounts. For instance, if the vote in a single district is 6,000 for the Darks and 4,000 for the Lights, then the Darks win. If 500 voters had switched from Dark to Light, Dark would still have won, but if 1,001 voters had switched from Dark to Light, Dark would have lost. If instead the votes had been 5,500 for Dark and 4,500 for Light, it would take only 501 voters to switch to affect the result. In short, the voting figures for a district don't just tell us who wins: they tell us how close the result was.

We can perform this calculation for each district, and combine the results to see how the number of seats won varies with the shift in votes, to get a seats–vote curve. (It's actually a polygon, with lots of straight edges, but it's convenient to smooth it.) The left-hand picture shows roughly how this curve ought to look for an election that's *not* been gerrymandered. In particular the curve ought to cross the 50% threshold for seats at the 50% split of votes, and it ought to be symmetric on either side of this point, when rotated 180°.

The right-hand picture shows the seats–vote curve for a map used for elections to Congress in Pennsylvania, with the Democratic vote along the horizontal axis. The Democrats would have had to win about 57% of the vote to secure 50% of the seats. This map was subsequently overturned by the state legislature.

In several cases, the US Supreme Court has rejected accusations of gerrymandering based on this kind of calculation, and it has also rejected claims based on lack of compactness of districts. In *LULAC v. Perry 2006* it did order a small number of district boundaries in Texas to be redrawn on the grounds that one district had been drawn contrary to the Voting Rights Act. In fact, although the Supreme Court has declared partisan gerrymandering to be unconstitutional, it has not yet struck down any complete districting map.

A major reason that the court gave for its negative decision

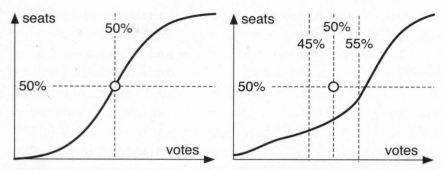

Graphing seats against votes. Horizontal axis shows percentage of votes for one party, and runs from 30% to 70%. Vertical axis shows percentage of seats that would have been won with that vote.

was that methods like the seats–vote curve rely on hypotheticals – what voters *would have done* in different circumstances. This might make sense to lawyers, but mathematically it's nonsense, because the curve is deduced from actual voting data by a precisely defined procedure. Transferring votes to calculate the curve doesn't depend on what any voter might have done in reality. It's like looking at a basketball score of 101–97 and deciding the game must have been close, while a score of 120–45 means it wasn't. You're not making predictions about what individual players might have done if they'd played better or worse. So we can add this one to the long and undistinguished list of the law's inability to grasp, or even appreciate, basic mathematics. The allegedly hypothetical nature of this entirely factual algorithm did, of course, provide the perfect excuse for not overturning the entire Texas map.

*

The best way to tackle questionable legal decisions is not to try to educate judges, so those seeking mathematical methods to detect gerrymandering looked for other measures that couldn't be challenged on spurious grounds. Gerrymandering forces the supporters of one party to waste many of their votes. Once your candidate has a majority, extra votes have no effect on the outcome. So one way to quantify the fairness, or not, of a boundary choice is to require both parties to waste roughly the same number of votes. In 2015 Nicholas Stephanopoulos and Eric McGhee defined one method to measure wasted votes, the efficiency gap.[12] In *Gill v. Whitford 2016*, a court in Wisconsin declared the state assembly map illegal, and the efficiency gap was instrumental in that decision. To see how to calculate the efficiency gap, let's simplify to elections with just two candidates.

There are two main ways to waste your vote. A vote cast for a losing candidate is wasted because you might as well not have bothered. An excess vote cast for the winner, after they get to 50%, is wasted for the same reason. These statements depend on the actual result and are applied with hindsight: you can't be sure your vote would have been wasted until you know the result. In the UK's 2020 general election the Labour candidate in my constituency got 19,544 votes, and the Conservative candidate got 19,143. Labour won by 401 votes out of a total of 38,687 votes cast for those two parties. If any single voter decided not to bother, the majority would still have been 400. But if just over 1% of Labour voters had decided not to bother, the Conservative candidate would have won.

According to the definition of wasted votes, the Conservative voters wasted a total of 19,143 votes and Labour voters wasted 200 votes. The efficiency gap measures the extent to which one party is forced to waste more votes than the other. In this case, it is:

Number of wasted Conservative votes
minus
Number of wasted Labour votes
divided by
Total number of votes.

That is, $(19143 - 200)/38687$, which is $+49\%$.

This is just one constituency. The idea is to calculate the efficiency gap for all constituencies combined, and get the lawmakers to set a legal target. The efficiency gap always lies between -50% and $+50\%$, and a gap of zero is fair since then both parties waste the same number of votes, so Stephanopoulos and McGhee suggested that an efficiency gap outside the range $\pm8\%$ is indicative of gerrymandering.

There are some flaws in this measure, however. When the result is close, a large efficiency gap is inevitable, and a few votes can swing it from near $+50\%$ to near -50%. My constituency wasn't gerrymandered, despite the $+49\%$ efficiency gap. If only 201 Labour voters had voted Conservative instead, it would have been -49%. If one party just gets lucky and wins every district, it will appear to have won through gerrymandering; demographic factors can distort the figures. In *Gill v. Whitford* the defence correctly pointed out these flaws, but the plaintiffs successfully argued that they didn't apply in that particular case. As general comments, however, they're entirely sensible.

In 2015 Mira Bernstein and Moon Duchin[13] spotted some other flaws in the efficiency gap, and in 2018 Jeffrey Barton suggested an improvement to eliminate them.[14] For example, suppose that there are eight districts, and in each district Light gets 90 votes while Dark gets the remaining 10. Light wastes $40 \times 8 = 320$ votes, whereas Dark wastes $10 \times 8 = 80$, so the efficiency gap is $(320 - 80)/800 = 0{\cdot}3 = 30\%$. If we follow the suggestion of an 8% threshold, this size of efficiency gap indicates partisan bias *against* Light. But Light won all of the eight seats!

A second scenario reveals another issue. Suppose now that Light wins three districts 51 to 49, while Dark wins two districts 51 to 49. Now Light wastes $1 + 1 + 1 + 49 + 49 = 101$ votes and Dark wastes $49 + 49 + 49 + 1 + 1 = 149$. The efficiency gap is $(101 - 149)/500 = -0{\cdot}096 = -9{\cdot}6\%$, indicating bias against Dark. However, Dark is the minority party and ought not to expect to

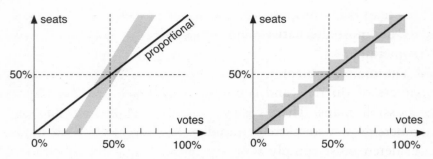

Left: Seats–votes graph showing proportional representation (thick line) and the region in which the efficiency gap is deemed to be fair (shaded). *Right*: Corresponding graph for the modified efficiency gap: the shaded region surrounds the diagonal line.

win more than two seats, as they do. Awarding Dark another seat would give the minority party a majority of seats.

Barton traces both issues to the uses of raw wasted votes. In *any* election, surplus votes for the winner are wasted, no matter how the boundaries are drawn. He replaces 'wasted votes' by '*unnecessarily* wasted votes', calculating the proportions of votes for each party that are bound to be wasted and subtracting them from the previous definition of wasted votes. With the original definition, the seats–vote graph gives a narrow band around the line from 25% of votes at the bottom to 75% at the top, as in the left-hand picture. For comparison, the diagonal line shows the ideal graph for proportional representation. The two coincide only very close to a 50/50 vote split. With unnecessarily wasted votes, the corresponding graph is shown at the right. It closely surrounds the diagonal, which is far more reasonable.

*

A different method for detecting gerrymandering is to consider alternative maps and compare the likely results, using data on probable voting patterns across the entire region being districted.

If the map being proposed by the Darks gives them 70% of the seats, but most alternative maps only given them 45%, they're up to something.

The main problem with this idea is that even for realistic numbers of districts and subdivisions, it's not possible to list all possible maps. There's a combinatorial explosion, and the numbers grow with extreme rapidity. Moreover, all of the maps considered must comply with the law, introducing constraints that can be mathematically intractable. As it happens, mathematicians long ago found a way to get round this combinatorial explosion: Markov Chain Monte Carlo (MCMC). Instead of surveying every possible map, MCMC creates a random sample of maps, big enough to provide accurate estimates. It's similar to the way public opinion polls estimate voters' intentions by asking a relatively small random sample.

Monte Carlo methods go back to the wartime Manhattan Project to build an atom bomb. A mathematician named Stanislaw Ulam had been ill, and was convalescing. To pass the time he played games of patience. Wondering what the chance of success was, he tried to estimate how many orderings of the pack of cards would lead to success with perfect play, and quickly realised this approach was hopeless. Instead, he played a lot of games and counted how often he won. Then he realised that he could play a similar trick with the physics equations that the Manhattan Project needed to solve.

Markov chains, named after the Russian mathematician Andrey Markov, are generalisations of the random (or drunkard's) walk. Someone a little the worse for wear has overindulged, and is staggering along the pavement, taking steps forward or back at random. On average, after a given number of steps, how far do they get? (Answer: on average, about the square root of the number of steps.) Markov imagined a similar process where the pavement is replaced by a network, and transitions along the edges of the network are assigned probabilities. A key issue is:

after wandering around for a very long time, what's the probability of being at any given location? Markov chains model many real-world problems in which sequences of events occur, with probabilities that depend on the current circumstances.

MCMC is what you get by putting the two together: use Monte Carlo methods to sample the required list of probabilities. In 2009 statistician Persi Diaconis estimated that about 15% of statistical analyses in science, engineering, and business are driven by MCMC, so it makes sense to try out such a powerful, established, and useful method on gerrymandering. Use random walks *à la* Markov to generate districting maps, sample them using Monte Carlo, and *bingo!* you have a statistical method for assessing how typical a proposed map is. There's some more sophisticated mathematics that supports these methods, known as ergodic theory, which provides a guarantee that sufficiently long random walks sample the statistics accurately.

Mathematicians have recently testified about MCMC before the courts. In North Carolina, Jonathan Mattingly compared MCMC estimates of the reasonable range of quantities such as seats won to argue that the plan that had been chosen was an extreme statistical outlier, indicating that it was partisan. In Pennsylvania, Wesley Pegden used statistical methods to calculate how unlikely it would be for a politically neutral plan to produce worse results than plans created by a random walk, and to estimate the likelihood of such a result occurring by pure chance. In both cases, the judges found the mathematical evidence credible.

*

Mathematical understanding of gerrymandering cuts both ways. It can help voters and law courts spot when it's happening, but it can also suggest more effective ways to gerrymander. It helps us make people stay within the law, but it can also help them to break it, or, perhaps worse, bend it. Whenever technical regulations are drawn up to prevent some kind of abuse, people game the system and scrutinise the regulations for loopholes. The great virtue of a mathematical approach is that it makes the rules themselves clear. It also raises an entirely new possibility. Instead of futile attempts to persuade competing political interests to agree on what's fair, giving them opportunities to game the system, and policing the system through the courts, it might be more sensible to let them *fight it out*. Not in a free-for-all, where power and money offer huge advantages, but in a framework that's structured to ensure not only that the outcome is fair, and seen to be fair, but also that the parties involved can't avoid accepting that it's fair.

This may seem a lot to ask, but an entire field of mathematics devoted to this idea has recently flowered: the theory of fair division. And it tells us that carefully structured frameworks for negotiation can achieve what initially seems impossible.

The classic example, from which everything else flows, is two children arguing over a cake. The problem is to divide it between them, using a protocol – a set of rules specified in advance – that's provably fair. The classic solution is 'I cut, you choose'. Alice is told to cut the cake so that she considers both pieces to be of equal value. Then Bob is asked to choose one of the pieces. Bob ought not to have any objections since he makes the choice. He could have chosen the other piece instead. Alice ought not to have any objections either: if she thinks Bob chose the bigger piece, she should have cut the cake differently in the first place. If they're bothered about who goes first, toss a coin, but actually that's not needed.

Human nature being what it is, we can't be certain that the children will see it that way after the event. When I mentioned this method in an article, a reader wrote in to say that he'd tried it on his children, and Alice (not her real name) had promptly complained that Bob (not his) had the bigger piece. When her father pointed out that this was her fault for cutting badly, the news didn't go down terribly well – in her eyes it amounted to blaming the victim – so her father swapped the two pieces. Only to hear her wail: 'Bob's piece is *still* bigger than mine!' But this kind of protocol ought to satisfy politicians, or at least shut them up, and it certainly ought to be acceptable in a court of law. The judge just has to check that the protocol has been followed correctly.

The key feature of this kind of protocol is that instead of trying to eliminate Alice and Bob's mutual antagonism, we *use* it to arrive at a fair outcome. Don't ask them to play fair, don't tell them to cooperate, don't propose some artificial legal definition of what 'fair' means. Just let them *oppose* each other and play the game. Of course, Alice and Bob have to agree beforehand to play by those rules, but they're going to have to agree to something, and the rules are transparently fair, so they're likely to get short shrift if they don't comply.

An important feature of 'I cut, you choose' is that it doesn't involve some external assessment of what a piece of cake is worth. It uses the players' own subjective estimates of its value. They just need to be satisfied that their share is fair *by their own criteria*. In particular, they don't need to agree on the value of anything. In fact, fair division is easier if they don't. One wants the cherry, the other the icing, neither is bothered about the rest: job done.

When mathematicians and social scientists started taking this kind of problem seriously, remarkable hidden depths emerged. The first step forward came when they thought about how three people should share a cake. Not only is the simplest answer distinctly tricky to find, but there's a new twist. Alice, Bob, and Charlie can agree that the result is fair, in the sense that they've got at least one third of the cake by their own valuation, but Alice might still envy Bob because she thinks Bob's share is bigger than hers. Charlie's share must compensate for this, in Alice's eyes, by being smaller than hers, but there's nothing contradictory about that, because Bob and Charlie can have different ideas of what their pieces are worth *to them*. So it makes sense to seek a protocol that's not only fair, but envy-free. This can, in fact, be achieved.[15]

The 1990s saw major advances in our understanding of fair and envy-free division, beginning with an envy-free protocol for dividing among four people, found by Steven Brams and Alan Taylor.[16] The cake is of course just a metaphor for something of value that's capable of being divided. The theory deals with items that can be divided as finely as we wish (cake) or that come in discrete lumps (books, jewellery). This makes it applicable to real-world issues of fair division, and Brams and Taylor explained how to use such methods to resolve disputes in divorce settlements. Their Adjusted Winner protocol has three major advantages: it's equitable, envy-free, and efficient (or Pareto-optimal). That is, each party feels that its share is at least as big as the average share, they feel no wish to trade shares with anyone else, and no other division exists that's at least as good for everybody and better for somebody.

In a divorce negotiation, for instance, it might work like this. After a lifetime of collaborating on the exchange of cryptographic messages, Alice and Bob get fed up and decide on a divorce. Each of them is given 100 points, which they divide up by assigning a number of points to each object – the house, the TV, the cat. Initially, the objects are given to whoever assigned them the most points. This is efficient, but usually it's neither fair nor envy-free, so the protocol moves to the next stage. If both scores are the same, everyone is satisfied and the division ends. If not, suppose that Alice's share, according to her points score, is bigger than Bob's share, according to his. Now objects are transferred from Alice (the *winner*) to Bob (the *loser*) in an order that ensures that both scores become equal. Because both the valuations and the objects are discrete, one of the objects might have to be subdivided, but the protocol implies that this will happen for at most one object – in all likelihood the house, which will be sold and the money then split, but not if Bob bought shares in Apple before it took off on the stock market.

Adjusted Winner satisfies three important conditions for fair division. It has a fairness guarantee: it's provably equitable, envy-free, and efficient. It works through multilateral evaluation: the individuals' preferences are taken into account, and values of their shares are calculated using their own evaluations. Finally, it's procedurally fair: both parties can understand and verify the fairness guarantee for whatever solution is eventually arrived at, and if necessary a court can determine that it's fair.

*

In 2009 Zeph Landau, Oneil Reid, and Ilona Yershov suggested that a similar approach could eliminate the problem of gerrymandering.[17] A protocol that stops any participant from drawing districts to their own advantage halts the gerrymander in its reptilian tracks. This method avoids any consideration of the shape of the maps, and it doesn't give allegedly unbiased outsiders the power to impose maps. Instead, it's set up so that competing interests balance each other.

Better still, these methods can be beefed up to take into account additional factors such as geographical cohesion and compactness. If an outside body such as an electoral commission is required to make the final decision, the results of the division game can be presented to it as part of the evidence on which it will base its judgement. No one is claiming that in the real world such methods remove every trace of bias, but they work a lot better than existing methods, and they largely remove the temptation to indulge in blatantly unfair practices.

The protocol, too complicated to describe in detail, involves an independent agent who proposes a way to split up the state. The parties are then offered the option of changing the agent's map by subdividing one of the pieces, provided that the other party is allowed to subdivide the other one. Alternatively, they can choose a similar option with the parties' roles reversed. This is a version of 'I cut, you choose' with more complicated sequences of cuts. Landau, Reid, and Yershov prove that their protocol is fair, from the point of view of either party. In essence, the two parties play a game against each other. But the game is designed to end in a draw, with each party convinced that it's done as well as it possibly could. If not, it should have played the game better.

In 2017 Ariel Procaccia and Pegden improved on this protocol by eliminating the independent agent, so that everything is decided by the two opposing parties. In outline, one political party divides a map of the state into the legally required number of districts, with (as near as possible) equal numbers of voters in each. Then the second party 'freezes' one district so no further changes can be made to it, and remaps the remaining districts, however it wants. The first party then chooses a second district from this new map, freezes it, and redraws the rest. The parties take turns freezing and redrawing, until everything is frozen. That decides the final map that will be used for districting. If there are, say, 20 districts, this process goes through 19 cycles. Pegden, Procaccia, and visiting computer science student Dingli Yu proved mathematically that this protocol doesn't give the first player any advantage, and that neither player can concentrate specific populations of voters into the same district if the other player doesn't want this to happen.

*

The mathematics of elections is now a very extensive subject, and gerrymandering is only one aspect. A lot of work has been done on different voting systems – first past the post, single transferable vote, proportional representation, and so on. One of the general themes that emerges from this research is that if you write down a short list of properties that are desirable in any sensible democratic system, it turns out that in some circumstances these requirements contradict each other.

The great-grandmother of all these results is Arrow's Impossibility Theorem, which the economist Kenneth Arrow published in 1950 and explained in his book *Social Choice and Individual Values* a year later. Arrow considered a ranked voting system, in which each voter assigns numerical scores to a series of options: 1 for their first preference, 2 for the next, and so on. He stated three criteria for fairness of such a voting system:

- If *every* voter prefers one alternative to another, then so does the group.
- If no voter's preference between two specific options changes, neither does that of the group, even if preferences between other options do change.
- There is no dictator who can always determine which option the group prefers.

All very desirable but, as Arrow proceeded to prove, logically contradictory. That doesn't mean that such a system is necessarily unfair: just that in some circumstances the outcome is counterintuitive.

Gerrymandering has its own descendants of Arrow's Theorem. One of them, published by Boris Alexeev and Dustin Mixon[18] in 2018, lays down three principles for fair districting:

- *One person, one vote*: each district has roughly the same number of voters.
- *Polsby–Popper compactness*: all districts have a Polsby–Popper score greater than some legally specified amount.
- *Bounded efficiency gap*: more technical. Roughly, if the populations of any two districts are at most some fixed proportion of the total population of those districts, then the efficiency gap is less than 50%.

They then prove that no districting system can always satisfy these three criteria.

Democracy can never be perfect. In fact, it's amazing it works at all, given that the aim is to persuade millions of people, each with their own opinion, to agree on something important that affects every one of them. Dictatorships are so much simpler. One dictator, one vote.

3

Let the Pigeon Drive the Bus

On the one hand, the bus driver might have been
concerned that the pigeon would not be able to safely
drive the bus. On the other hand, maybe the driver
was more concerned that the pigeon would not be able
to take a route that would efficiently pick up all the
passengers at the various stops throughout the city.

Brett Gibson, Matthew Wilkinson,
and Debbie Kelly, *Animal Cognition*

Mo Willems drew cartoons from the age of three. Worried that
adults might be praising him dishonestly, he started writing funny
stories. Fake laughs, he felt, would be easier to spot. In 1993 he
joined the writing and animating team for the iconic *Sesame
Street*, winning six Emmy awards in ten years. His children's
TV cartoon series *Sheep in the Big City* featured Sheep, whose
idyllic farm life is shattered when General Specific's secret military organisation wants him for its sheep-powered ray-gun. His
first foray into children's books continued the animal theme with
Don't Let the Pigeon Drive the Bus!, winning a Carnegie medal
for its animated adaptation, and a Caldecott Honor, which you
get if you're shortlisted for the Caldecott Medal. The main character – a pigeon, obvs – uses every trick in the book (literally) to
convince the reader that it should be allowed to drive a bus when
the regular human driver suddenly has to leave.

Willems's book had an unintended scientific consequence

in 2012, when the entirely respectable journal *Animal Cognition* published an entirely respectable paper by the entirely respectable researchers Brett Gibson, Matthew Wilkinson, and Debbie Kelly. They showed experimentally that pigeons can find solutions, close to optimal, to simple cases of a famous mathematical curiosity: the Travelling Salesman Problem. Their title was 'Let the pigeon drive the bus: pigeons can plan future routes in a room'.[19]

Let no one claim that scientists lack a sense of humour. Or that cute titles don't help to generate publicity.

The Travelling Salesman Problem is not just a curiosity. It's a very important example of a class of problems of enormous practical significance, called combinatorial optimisation. Mathematicians have a habit of posing deep and significant questions in terms of apparent trivia. US congressmen have denounced the waste of public money on knot theory, unaware that this area is central to low-dimensional topology, with applications to DNA and quantum theory. Basic techniques in topology include the hairy ball theorem and the ham sandwich theorem, so I suppose we ask for it, but it's not just us. I don't mind the ignorance – that can happen to anyone – but why don't these people just *ask*?[20]

Anyway, the piece of significant trivia that inspires this chapter had its origins in a helpful book for – you guessed it – travelling salesmen. Door-to-door sellers. I can remember them even if you don't. They often sold vacuum cleaners. Like any sensible business person, the German travelling salesman of 1832 (and in those days it always *was* a man) placed a premium on using his time efficiently and cutting costs. Fortunately, help was at hand, in the form of a manual: *Der Handlungsreisende – wie er sein soll und was er zu thun hat, um Aufträge zu erhalten und eines glücklichen Erfolgs in seinen Geschäften gewiss zu sein – von einem alten Commis-Voyageur* (The travelling salesman – how he should be and what he has to do, to obtain orders and to be sure of a happy success in his business – by an old travelling salesman). This elderly peripatetic vendor pointed out that:

> Business brings the travelling salesman now here, then there, and no travel routes can be properly indicated that are suitable for all cases occurring; but sometimes, by an appropriate choice and arrangement of the tour, so much time can be gained, that we do not think we may avoid giving some rules also on this … The main point always consists of visiting as many places as possible, without having to touch the same place twice.

The manual didn't propose any mathematics to solve this problem, but it did contain examples of five allegedly optimal tours through Germany (one passing through Switzerland). Most of them involve subtours that visit the same place twice, perfectly practical if you're staying overnight in an inn and spend the day visiting the local area. But one of them made no repeat visits. A modern solution of the same problem shows that the manual's answer is pretty good, as the picture shows.

The Travelling Salesman Problem, or TSP, as it came to be known – later changed to Travelling Salesperson Problem to avoid sexism, which conveniently has the same acronym – is a founding example for the mathematical area now known as combinatorial optimisation. Which means 'finding the best option among a range of possibilities that's far too big to check one at a time'. Curiously, the TSP name seems not to have been used explicitly in any publication concerning this problem until 1984, although it was common usage much earlier in informal discussions among mathematicians.

For a problem with such practical origins, the TSP has led the mathematical community into very deep waters indeed, including the Millennium Prize Problem 'P ≠ NP?', whose million-dollar prize still awaits a claimant. This asks, in a precise technical sense,

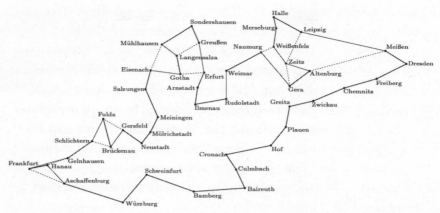

Tour (1,285 km) of 45 German cities from the 1832 manual, shown by the unbroken (bold and thin) lines. Solid bold and dashed lines show a shortest tour (1,248 km) found by modern methods.

whether, given a problem for which a proposed answer – a guess, if you wish – can be *verified* efficiently, you can always *find* the answer efficiently. Most mathematicians and computer scientists believe the answer is 'no': surely, checking any particular guess can be done much faster than finding the correct answer. After all, if someone shows you a solved 500-piece jigsaw puzzle, a quick once-over can usually tell if they got it right – but putting the puzzle together initially is another matter entirely. Unfortunately, jigsaw puzzles don't provide an answer: they're a useful metaphor, but technically they don't do the job. So right now, nobody can prove or disprove the belief that P is different from NP, which is why a solution will net you a cool million dollars.[21] I'll come back to P ≠ NP later, but first let's take a look at the early progress on solving the TSP.

*

The age of the travelling salesman is long gone, and that of the less sexist travelling salesperson quickly followed. In the age of the Internet, companies seldom sell their goods by sending someone from town to town with a suitcase full of samples. They put everything on the web. As usual (unreasonable effectiveness) this change of culture hasn't made the TSP obsolete. As online shopping grows exponentially, the demand for efficient ways to determine routes and schedules is becoming ever more important for everything from parcels to supermarket orders to pizza. The TSP should probably be renamed the Tesco Shopping Problem: what's the best route for a delivery van?

The portability of mathematics also comes into play. Applications of the TSP are not restricted to travel between towns or along city streets. On the wall of our sitting room is a large square of black cloth, embroidered in blue with an elegant pattern of spirals based on the famous Fibonacci numbers, picked out in sequins. The designer calls it the Fibonacci Sequins. It was made using a computer-controlled machine, capable of embroidering anything up to the size of a bedspread. The needle that sews the threads is attached to a rod, which it can slide along, and the rod can move perpendicular to its length. Combining the two motions, the needle can be moved anywhere you like. For practical reasons (wasted time, stress on the machine, noise) you don't want it hopping around all over the place, so the total distance should be minimised. That's very like the TSP. The ancestry of such machines goes back to the early days of computer graphics, and a gadget known as an XY plotter, which moved a pen in the same manner.

Similar issues abound in science. Once upon a time, prominent astronomers had their own telescopes, or shared them with a few colleagues. The telescopes could easily be redirected to point at new heavenly bodies, so it was easy to improvise. Not so any more, when the telescopes used by astronomers are enormous, ruinously expensive, and accessed online. Pointing the telescope at a fresh object takes time, and while the telescope is being moved, it can't be used for observations. Visit targets in the wrong order and a lot of time is wasted moving the telescope a long way, and then back again to somewhere near where it started. In DNA sequencing, fragmentary sequences of DNA bases must be joined together correctly, and the order in which this is done has to be optimised to avoid wasting computer time.

Other applications range from routing aircraft efficiently to the design and manufacture of computer microchips and printed circuit boards. Approximate solutions of TSPs have been used to find efficient routes for Meals on Wheels and to optimise the delivery of blood to hospitals. A version of the TSP even showed up in 'Star Wars', more properly President Ronald Reagan's hypothetical Strategic Defense Initiative, where a powerful laser orbiting the Earth would have been targeted at a series of incoming nuclear missiles.

*

Karl Menger, some of whose work is now seen as a forerunner of fractals, seems to have been the first mathematician to write about the TSP, which he did in 1930. He came to the problem from a very different angle: he'd been studying the lengths of curves from the viewpoint of pure mathematics. At the time, the length of a curve was defined as the largest value obtained by adding together the lengths of any polygonal approximation to the curve, whose vertices are a finite set of points on the curve that are visited in the same order as they lie on the curve. Menger proved that you get the same answer by replacing each polygon by a finite set of points on the curve, and finding the minimum total distance along *any* polygon with those vertices, in whichever order you wish. The connection with the TSP is that Menger's shortest path is the one that solves the TSP with the polygon's vertices considered as towns. Menger called it the 'messenger problem', arguing that it applied to postmen as well as salesmen, and wrote:

> This problem is solvable by finitely many trials. Rules which would push the number of trials below the number of

permutations of the given points, are not known. The rule
that one first should go from the starting point to the closest
point, then to the point closest to this, *etc.*, in general does
not yield the shortest route.

This quotation shows that he understood two key features of
the problem. First, there *exists* an algorithm to find the answer.
Just try all tours in turn, compute their lengths, and see which
is shortest. The total number of possible tours is precisely the
number of permutations of the points, which is finite. When he
wrote, no better algorithm was known, but trying every possibility is hopeless for more than a dozen or so towns, because there
are too many routes. Second, he knew that the 'obvious' method
– from each point go to the nearest one – usually doesn't work.
Experts call this method the 'nearest neighbour heuristic'. The
figure shows one reason why it fails.

Menger was a visiting lecturer at Harvard University for
six months from 1930 to 1931, and the great topologist Hassler
Whitney attended and made some suggestions about the problem.
A year later Whitney gave a talk in which he mentioned finding the
shortest route among the (then) 48 states of America. The name
'48 States Problem' circulated for a time, and no one seems to
know for certain who coined the snappier name 'Travelling Salesman Problem'. The first known reference in print using the TSP
name is a 1949 report by Julia Robinson.

Menger continued to work on the TSP and related issues. In
1940 László Fejes Tóth looked into essentially the same problem:
finding the shortest path through n points in the unit square. In
1951 Samuel Verblunsky proved that the answer has length less
than $2 + \sqrt{2 \cdot 8n}$. Various mathematicians proved slightly better
theorems showing that the minimal length over n points in a fixed
region is no more than some constant times the square root of n,
for constants that got smaller each time.

In the late 1940s one of the leading institutions for operations

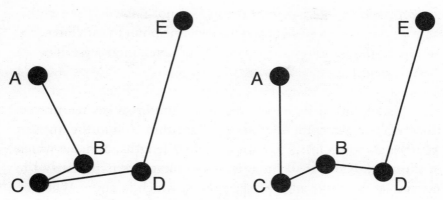

One way for the nearest neighbour heuristic to fail. Starting
from A and moving always to the nearest town among those still
unvisited, the left-hand tour visits ABCDE in turn. However,
the right-hand tour, which visits ACBDE, is shorter.

research was the RAND Corporation in Santa Monica, California.
The RAND researchers had been doing a lot of work on a related
question, the Transportation Problem, and George Dantzig and
Tjalling Koopmans suggested that their work on what's now
called linear programming might be relevant to the TSP. Linear
programming is a powerful and practical framework for many
combinatorial optimisation problems. It's a method for maximis-
ing some linear combination of variables, subject to inequalities
stating that certain other linear combinations must be positive
or negative. Dantzig invented the first practical algorithm, the
simplex method, still widely used. The inequalities define a multidimensional convex polyhedron, and the algorithm moves a point
along edges that increase the quantity we want to maximise, until
it gets stuck.

The first really significant progress on the TSP was made
in 1954 by RAND researchers Dantzig, Delbert Fulkerson, and
Selmer Johnson, using Dantzig's linear programming method.
They adapted it to apply to the TSP, and introduced systematic

new methods, in particular the use of 'cutting planes'. The upshot was a lower limit for the optimal tour length. If you can find a tour whose length is only slightly bigger, you're getting warm, in which case animal cunning can sometimes get you all the way. Dantzig, Fulkerson, and Johnson used these ideas to obtain the first solution to the TSP with a reasonable number of cities, namely the shortest tour through 49 cities: one in each of the 48 US states, plus Washington DC. This is probably the problem that Whitney mentioned in the 1930s, and exactly the problem mentioned by Robinson in 1949.

*

In 1956 operations research pioneer Merrill Flood argued that the TSP is likely to be hard. This raises a key question: how hard? To answer that, we must revisit P and NP, those million-dollar measures of computational complexity. It looks highly likely that Flood was right, in a very strong sense.

Mathematicians have always kept an eye on the practicality of methods for solving problems, although, when it comes to the crunch, they feel that any method is better than none. For purely theoretical purposes, merely being able to prove that a solution to some problem *exists* can be a major step forward. Why? Because if you can't be sure it exists, you could be wasting your time looking for it.

My favourite example here is what I call Mother Gnat's Tent. Baby Gnat is hovering a foot (metre, mile – anything bigger than zero) above the floor. Mother Gnat wants to make a tent with its base on the floor that will cover Baby Gnat, and she wants to use as little material as possible. Which tent has the smallest area? If we model Baby Gnat as a single point, the answer is 'there's no such thing'. You can make a tall thin conical tent with any area that's bigger than zero, but a tent with surface area zero is a line, not a tent. Given any tent, there's another one that does the job using half as much material. So there can't be a smallest area.

For the TSP, with any finite set of towns, arranged however we like, a solution definitely exists, because there are finitely many possible tours. This guarantees that you're not wasting your time trying to find the shortest route, but it doesn't tell you what it is. If you're hunting for buried treasure, it's no great help to be told that it's definitely *somewhere*: digging up the entire planet isn't practical.

Computer scientist Donald Knuth remarked, long ago, that in computing you need more than a proof that the answer exists. You need to find out how much it will cost to calculate it. Not in dollars and cents, but in computational effort. The mathematical area that addresses this issue is called computational complexity theory. In a very short time it has gone from a few simple ideas to a sophisticated collection of theorems and methods, but one basic distinction goes some way towards capturing, in very simple terms, the difference between a practical solution and an impractical one.

The main issue is: how fast does the running time (measured as the number of calculation steps) in any method for calculating the answer to a problem grow compared to the size of data needed to state the problem in the first place? Specifically, if it takes n binary digits to specify the problem, how does the running time depend on n? For practical algorithms, the running time tends to grow like a power of n, say n^2 or n^3. These algorithms are said to run in polynomial time, symbolised as class P. Impractical algorithms grow much faster, often in exponential time, like 2^n or 10^n. The 'try all tours' algorithm for the TSP is like this; it runs in factorial time $n!$, which grows faster than any exponential. In between is a grey area where the running time is bigger than any polynomial, but smaller than exponential. Sometimes these algorithms are practical, sometimes not. For present purposes we can take a very strict view and consign them all to a dustbin marked 'not-P'.

This is not the same as NP.

That acronym, confusingly, stands for a subtler idea altogether:

nondeterministic polynomial time. This refers to the running time of an algorithm that can decide whether any particular proposed solution is correct. Recall that a number is *prime* if it has no divisors except 1 and itself, so 2, 3, 5, 7, 11, 13, and so on, are prime. It is composite otherwise. So 26 is composite, since it equals 2 × 13. The numbers 2 and 13 are the prime factors of 26. Suppose that you want to find a prime factor of a number with 200 decimal digits. You spend a year fruitlessly seeking one, and in despair you consult the Delphic Oracle. It tells you that some particular large number is the answer. You have no idea where this came from (it is, after all, an oracle with miraculous powers of divination), but you can sit down and do the sums to see whether the oracle's number does indeed divide the very big number under consideration. Such a calculation is much, much easier than finding the prime factor itself.

Suppose that whenever the oracle proposes an answer, you can check whether it's correct using a polynomial-time (P) algorithm. Then the problem itself is class NP – nondeterministic polynomial. The oracle has a far harder task than yours, but you can always decide whether it told you the right answer.

It stands to reason that checking a proposed answer ought to be a lot easier than finding it. Checking a spot marked X for buried treasure is a lot easier than finding out where X is to begin with. For a mathematical example, almost everyone believes that finding the prime factors of a number is much harder than checking that a given prime *is* a factor. The main evidence is that fast algorithms are known for checking any proposed factor, but not for finding one. If P = NP, then given any problem that has a rapidly *checkable* answer, it would be possible to *find* the answer rapidly too. That sounds far too good to be true, and mathematicians' experience of solving problems is quite the opposite. So almost everyone believes that P ≠ NP.

However, all attempts to prove this, or disprove it, got stuck. You can prove a problem is NP by writing down an explicit algorithm and calculating its running time, but to prove it's *not* in P you have to consider *all possible algorithms* for solving it, and show that none of them is in class P. How can you do that? Nobody has a clue.

One curious fact that emerges from these attempts is that an awful lot of candidate problems are on the same footing. All of these problems are NP. Moreover, if any particular one can be proved not to lie in P, then none of them lies in P. They live or die together. Problems like this are said to be NP-complete. A related larger category is NP-hard: this consists of algorithms that can simulate the solution of *any* NP problem in polynomial time. If this algorithm turns out to have polynomial running time, this automatically proves the same is true for any NP problem. In 1979, Michael Garey and David Johnson proved that the TSP is NP-hard.[22] Assuming that P ≠ NP, this implies that any algorithm to solve it has running time larger than any polynomial.

Flood was right.

*

That's not a good reason to give up altogether, because there are at least two potential ways forward.

One, which I'll explore straight away, is based on experience of practical problems. If a problem is not-P, then solving it in the worst-case scenario is hopeless. But worst-case scenarios often turn out to be very contrived, and not typical of the examples you run into in the real world. So mathematicians in operations research set out to see just how many towns they could handle for real-world problems. And it turned out that variations on the linear programming method proposed by Dantzig, Fulkerson, and Johnson often perform remarkably well.

In 1980 the record was 318 cities; by 1987 it was 2,392 cities. By 1994 the record had risen to 7,397 cities, an answer that took about three years of CPU time on a network of very powerful computers. In 2001 an exact solution for 15,112 German towns was obtained using a network of 110 processors. It would have taken more than twenty years on a normal desktop. In 2004, the TSP was solved for a tour of all 24,978 towns in Sweden. In 2005, the Concorde TSP Solver solved the TSP for a tour of all 33,810 points on a printed circuit board. Setting records isn't the only reason for such research: the methods used to set them work very fast indeed for smaller problems. Up to a hundred cities can usually be solved in a few minutes, and up to a thousand in a few hours on a standard desktop machine.

The other option is to settle for less: a solution that's not too far from the best possible, but easier to find. In some cases, this can be achieved using a startling discovery made in 1890, in an area of mathematics so novel that many of the leading figures at the time failed to see any value in it, and often failed to believe the answers that more visionary mathematicians were slowly finding. Worse, the problems they tackled seemed to be 'mathematics for its own sake', bearing no visible relationship to anything in the real world. Their results were widely considered to be highly arti-ficial and the new geometric shapes that they constructed were dubbed 'pathological'. Many felt that even if those results were correct, they didn't advance the cause of mathematics one iota; they just threw up silly obstacles to progress in a self-indulgent orgy of logical nitpicking.

*

One method for finding good, but less than optimal, solutions to the TSP emerged from one of those silly obstacles. For a few decades either side of 1900, mathematics was in transition. The earlier buccaneering spirit of daring advances that ignored awkward details had just about run its course, and its disregard of basic issues such as 'what are we really talking about here?' or 'is this actually as obvious as we all think?' were sowing confusion and perplexity where there ought to have been clarity and insight. Worries about advanced areas like calculus, where mathematicians had been flinging infinite processes around with gay abandon, were slowly working their way backwards from the esoteric to the commonplace. Instead of having doubts about integrals of complicated mathematical functions like the complex logarithm, people were wondering what a function was. Instead of defining a curve to be continuous if it could be 'freely drawn with the hand', they sought greater rigour, and found it lacking. Even the nature of something as basic and obvious as a number was proving to be elusive. Not just for novel constructs such as complex numbers: for good old whole numbers 1, 2, 3. Mainstream mathematics continued to advance, tacitly assuming that issues of this kind would eventually be sorted out and all would be well. The logical status of the foundations could safely be left to the hairsplitters and pedants. And yet … a general feeling that this cavalier approach to the subject could not go on much longer began to crystallise.

Things really started to go wrong when the older swashbuckling methods began turning up answers that contradicted each other. Theorems long believed to be true turned out to be false in exceptional, usually rather strange, circumstances. An integral, calculated in two different ways, gave two different answers. A series, believed to converge for all values of the variable, sometimes diverged. It wasn't as bad as finding that $2 + 2$ is sometimes 5, but it did make some people wonder what 2 and 5 really were, not to mention + and =.

So, undeterred by the majority naysayers – or, at least, not deterred enough to change their minds – a few nitpickers burrowed down through the mathematical edifice, from the lofty heights to the basement beneath, in search of solid ground, and then began to renovate the building from bottom to top.

Like all renovations, the eventual outcome differed from the original in subtle but disturbing ways. The notion of a curve in the plane, which had been around since the time of the ancient Greeks, had hidden depths. The traditional examples – the circles, ellipses, and parabolas of Euclid and Eratosthenes, the quadratrix that the Greeks used to trisect angles and square the circle, the figure-eight lemniscate of the Neoplatonist philosopher Proclus, the ovals of Giovanni Domenico Cassini, the cycloids and more complicated offspring such as Ole Rømer's hypocycloids and hypercycloids – held their own fascination, and had led to remarkable advances. But, just as domesticated animals give a misleading picture of life in the Earth's rainforests and desert wildernesses, these curves were much too tame to represent the wild creatures that roamed the mathematical jungle. As examples of the potential complexity of continuous curves, they were too simple and too well behaved.

One of the most basic features of curves, so obvious that no one sought to question it, is that they're *thin*. As Euclid wrote in his *Elements*, 'a line is that which has no thickness'. The area of a line – just of the line, not whatever it encloses – is evidently zero. But in 1890, Giuseppe Peano gave a construction for a continuous curve that completely fills the interior of a square.[23] It doesn't just wander around inside the square in a complicated scribble that comes close to any point: it passes though *every* point in the square, hitting it exactly. Peano's curve does indeed 'have no thickness', in the sense that you make it by tracing a line with a pencil whose tip is a single geometric point, but that line wiggles around in a very convoluted manner, repeatedly revisiting regions that it has previously left. Peano realised that if you make it infinitely wiggly, in a carefully controlled manner, it will fill the entire square. In particular, the area of the curve is the same as that of the square – so it's not zero.

This discovery came as a shock to naive intuition. At the time, curves of this type were called 'pathological', and many mathematicians reacted to them the way we usually react to pathology – with fear and loathing. Later, the profession got used to them and absorbed the deep topological lessons that they teach us.

Today we see Peano's curve as an early example of fractal geometry, and we appreciate that fractals are in no way unusual or pathological. They're commonplace, even in mathematics, and in the real world they provide excellent models of highly complex structures in nature, such as clouds, mountains, and coastlines.

The pioneers of this new age of mathematics inspected ancient intuitive concepts like continuity and dimension, and started asking the difficult questions. Instead of assuming that they could get away with the traditional tricks used in simpler areas of mathematics, these pioneers asked whether those tricks work in sufficient generality, and if so, *why* they work. Or, if they don't always work, what goes wrong. This sceptical approach annoyed many mainstream mathematicians, who saw it as negativity for its own sake. 'I turn away with fright and horror of this terrible scourge of continuous functions without derivative,' Charles Hermite wrote in 1893 to his friend Thomas Stieltjes.

The traditionalists were much more interested in pushing out the boundaries by assuming everything in the logical garden was lovely, but the new scepticism, with its flurry of bizarre challenges to intuition, was a necessary reaction against naivety. By the 1930s, the value of this more rigorous approach was becoming evident; by the 1960s, it had taken over almost completely. You could write an entire book about this period of our subject's development, and some authors have done just that. Here, I want to concentrate on one subtheme: continuous curves and the concept of dimension.

*

The concept of a curve probably goes back to the time when some early human first dragged the tip of a stick across a patch of sand or mud and discovered that it left a trail. It began to acquire its current form when a logical approach to geometry got off the ground in ancient Greece, and Euclid asserted that a point has only position and a line has no thickness. A curve is a line that need not be straight, the simplest example being a circle, or an arc thereof. The Greeks developed and analysed a variety of curves – the aforementioned ellipse, quadratrix, cycloid, and so on. They discussed only specific examples, but it was 'sort of obvious' how the general idea should go.

After the introduction of calculus, two properties of curves came to the fore. One was continuity: a curve is continuous if it has no breaks. The other, more delicate, was smoothness: a curve is smooth if it has no sharp corners. Integral calculus works best for continuous curves, and differential calculus works best for smooth ones. (I'm being *very* sloppy here, to keep the story moving, but I claim I'm closer to the truth than to fake news.) Of course, it wasn't quite that simple: you had to define 'break' and 'corner', and do so *precisely*. More subtly, whatever definitions you came up with had to be suitable for mathematical study, phrased in mathematical terms. You had to be able to *use* them. The details still baffle mathematics undergraduates the first time they encounter them, so I'll spare you those.

The second key concept is dimension. We all learn that space has three dimensions, a plane has two, and a line has one. We don't approach this idea by defining the word 'dimension' and then counting how many of them space, or a plane, has. Not exactly. Instead, we say that space has three dimensions because we can specify the position of any point using exactly three numbers. We choose some specific point, the origin, and three directions: north–south, east–west, and up–down. Then we just have to measure how far it is from the origin to our chosen point, in each of those directions. This gives us three numbers (the *coordinates* relative to those choices of direction), and each point in space corresponds to one, and only one, such triple of numbers. Similarly, a plane has two dimensions because we can dispense with one of those numbers, say the up–down one, and a line has one dimension.

That all seems pretty easy until you start thinking. The

previous paragraph assumes that the plane in question is horizontal. That's why up–down can be thrown out. But what if the plane is on a slope? Then up–down matters. However, it turns out that the up–down number is always determined by the other two (provided you know how steep the slope is). So what matters isn't the number of directions along which you measure coordinates: it's the number of *independent* directions. That is, directions that aren't combinations of other directions.

It's getting a bit more complicated now, because we can't just count how many coordinates there are. It's more a case of the smallest number that will do the job. And that raises another, rather deeper, question: how do you know that two actually *is* the smallest number that will do the job for a plane? It might well be true – and if not, we need a better definition – but it's not totally obvious. Now the floodgates open. How do we know that three is the smallest number that will do the job for space? How do we know that *any* choice of independent directions always gives three numbers? For that matter, how sure are we that three numbers are enough?

That third question is really one for experimental physics, and leads, via Einstein and his General Theory of Relativity, to the suggestion that physical space is not, in fact, the flat three-dimensional space of Euclid, but a curved version. Or, if the string theorists are correct, spacetime has ten or eleven dimensions, all but four of which are either too small for us to notice, or inaccessible. The first and second questions can be resolved satisfactorily, but not trivially, by defining three-dimensional Euclidean space in terms of a coordinate system with three numbers, and then spending five or six weeks of a university course on vector spaces, where any number of coordinates is possible, to prove that the dimension of a vector space is unique.

Inherent in the vector space approach is the idea that our coordinate system is based on straight lines, and the space is flat. Indeed, another name is 'linear algebra'. What if we do an

Einstein and allow the coordinate system to bend? Well, if it bends smoothly (classically called 'curvilinear coordinates') all is well. But in 1890 the Italian mathematician Giuseppe Peano discovered that if it bends in a wild manner – so wild that it's no longer smooth, but remains continuous – then a space of two dimensions can have a coordinate system with only *one* number. The same goes for a space of three dimensions. In this more general, flexible set-up, suddenly 'the' number of dimensions becomes mutable.

One response to this strange discovery is to dismiss it; obviously we have to use smooth coordinates, or whatever. But it turned out to be much more creative, and useful, and indeed more *fun*, to embrace the weirdness and see what happens. The traditionalist critics were rather puritanical, and they didn't want the younger generation to have any fun at all.

*

Down to brass tacks. What Peano discovered – constructed – was a continuous curve that passes through every point in a square. Not just its boundary, that's easy: the entire interior, too. And the curve really must hit every point exactly, not just come very close.

Suppose such a curve exists. Then it's just some kind of wiggly line, which has its own intrinsic coordinate system – how far along the line we have to go. That's one number, so the curve is one-dimensional. However, if this wiggly curve passes through every point of a solid square, which is two-dimensional, we have now managed to specify every point of that square using just one continuously varying number. So the square is actually one-dimensional!

I generally avoid exclamation marks when writing, but this discovery deserves one. It's crazy. Also true.

Peano had found the first example of what we now call a 'space-filling' curve. Its existence relies on the subtle but vital distinction between smooth curves and continuous ones. Continuous curves can be wiggly. Smooth ones … can't. Not *that* wiggly.

Peano had the right sort of mind to invent his curve. He liked the fine points of logical detail. He was also the first person to write down precise axioms for the system of whole numbers – a simple list of properties that specify that system precisely. He didn't invent his space-filling curve just for the fun of it: he was putting the finishing touches to the work of a similarly-minded predecessor, who also had a deep interest in the nature of whole numbers and counting. His name was Georg Cantor, and what really interested him was infinity. Most leading mathematicians of the period rejected Cantor's radical and brilliant ideas, driving him to despair. Rejection probably wasn't the cause of his later mental illness, as is sometimes suggested, but it sure didn't help. Among the few top mathematicians who appreciated what Cantor was trying to do was one who scaled the mathematical heights to the topmost pinnacle: David Hilbert. Hilbert was perhaps the leading mathematician of his age, and in later life he became another of those pioneers of mathematical logic and the foundations of the subject. Maybe he recognised a fellow spirit.

At any rate, it all began with Cantor, and his introduction of transfinite cardinals – how to count how many members an infinite set has. He famously proved that some infinities are bigger than others. More precisely, there's no one-to-one correspondence between the integers and the real numbers. Seeking a transfinite cardinal larger than that of the reals, for a time he became convinced that the cardinal of the plane must be greater than that of a line. In 1874 he wrote to Richard Dedekind:

Can a surface (say a square that includes the boundary) be uniquely referred to a line (say a straight line segment that includes the end points) so that for every point on the surface there is a corresponding point of the line and, conversely, for every point of the line there is a corresponding point of the surface? I think that answering this question would be no easy job, despite the fact that the answer seems so clearly to be 'no' that proof appears almost unnecessary.

Three years later he wrote again to say he'd been wrong. Very wrong. He'd found a one-to-one correspondence between the unit interval and n-dimensional space for any finite n. That is, a way to match members of the sets so that each member of one matches exactly one of the other. 'I see it,' Cantor wrote, 'but I don't believe it!'

The main idea is simple: given two points in the unit interval (between 0 and 1) we can write them in decimals as

$$x = 0 \cdot x_1 x_2 x_3 x_4 \ldots$$
$$y = 0 \cdot y_1 y_2 y_3 y_4 \ldots$$

and let this correspond to a point in the unit interval whose decimal expansion is

$$0 \cdot x_1 y_1 x_2 y_2 x_3 y_3 x_4 y_4 \ldots$$

by interleaving the decimal places, like a riffle shuffle interleaves two halves of a pack of cards.[24] The main difference is that Cantor's pack of cards is infinite. Now, when you riffle shuffle two infinite packs together, you get *one* infinite pack. This is how Cantor manages to fit two coordinates into one. To handle three dimensions, just use three packs, and so on.

Cantor published some of these results in 1878. He investigated countable sets, which can be placed in one-to-one correspondence with the counting numbers, and sets that are in one-to-one correspondence with each other. He also realised that his correspondence between the unit interval and the unit square doesn't preserve dimension – one dimension goes to two – and, crucially for our story, he emphasised that the correspondence that he'd constructed isn't continuous. That is, points very close together in the unit interval need not correspond to points very close together in the unit square.

Cantor's ideas were controversial: some eminent mathematicians considered them nonsense, probably because they were so original that it required imagination and an open mind to appreciate them. Others, notably Hilbert, declared the new area that Cantor had opened up to be a 'paradise'. Full recognition of the importance of Cantor's work came only after his death.

*

In 1879 Eugen Netto[25] answered one obvious question by proving that there's no *continuous* one-to-one correspondence between the unit interval and the solid unit square, which is trickier than it might seem. The most significant breakthrough came in 1890, when Peano set the cat among the pigeons with his space-filling curve, showing that our default mental image of a continuous curve can be distinctly misleading.

Peano's paper has no pictures. He defines the curve using base-3 expansions of points in the unit interval, and his construction is equivalent to the geometric one in the left-hand picture overleaf.[26] In 1891 Hilbert published another example of a space-filling curve, drawing a picture like the one on the right. Both constructions are fairly complicated: the pictures show an early stage of a recursive process in which simple polygons are repeatedly replaced by more elaborate ones. Many other space-filling curves have since been found.

Space-filling curves have applications to computing, such as storage and retrieval of multidimensional data.[27] The basic idea is that we can traverse a multidimensional array by following an approximation to a space-filling curve, reducing the problems to the one-dimensional case. Another application yields a quick-and-dirty solution of the Travelling Salesperson Problem. The idea is

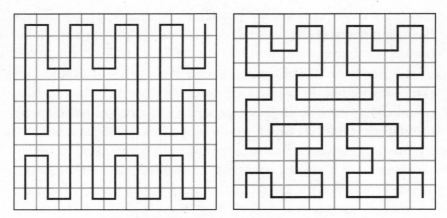

Left: An early stage in a geometric interpretation of
Peano's space-filling curve. *Right*: An early stage in the
construction of Hilbert's space-filling curve.

to run a finite approximation to a space-filling curve through the
region containing the cities, put the cities in order along the curve,
and then visit them in that order using the shortest linking route
at each step. This produces a route that's usually no more than
25% longer than the optimal one.[28]

What other shapes can a curve fill? Hilbert's construction
extends to three dimensions, giving a curve that fills the unit cube,
and curves can also fill hypercubes of any dimension. The last
word is a theorem proved by Hans Hahn and Stefan Mazurkiewicz, which completely characterises the topological spaces that
a curve can fill.[29] Almost anything, it turns out, provided it's
compact (finite in extent) and satisfies a few technical conditions
to rule out silly spaces.

*

The travelling salesperson may yet have the last word. In 1992
Sanjeev Arora and coworkers[30] discovered that the complexity

class NP ('easily checkable') has a curious property that casts doubt on the prospects for finding class-P algorithms ('easily calculated') that give good approximate solutions. They proved that if P ≠ NP, and the size of the problem is above some threshold, it's no easier to compute a good approximation to the answer than to find the answer itself. The only alternative to this conclusion would be that P = NP, which would win a million dollars, but has to remain hypothetical.

Their work is related to a truly remarkable idea: transparent proofs. Proofs are the essence of genuine mathematics. In most branches of science you can test your theories against reality by making observations or performing experiments. Mathematics lacks that luxury, but it still has a way to verify its results. First, they must be supported by a logical proof. Second, that proof must be checked to make sure there are no errors and no loopholes. This ideal is hard to achieve, and it's not quite what mathematicians actually do, but it's what they aim at. Anything that fails such a test is immediately labelled 'wrong', though it may still be useful as a step towards a better proof that's right. So, from the time of Euclid until the present day, mathematicians have spent a lot of time going carefully over proofs, their own or others', line by line, looking for things they agree with and things that don't quite stack up.

In recent years a different way to verify proofs has appeared: use a computer. This requires rewriting proofs in a language that computers can process algorithmically. It works, and it's had some serious successes on some of the hardest proofs in the journals, but so far it's not displaced more traditional methods. A side effect of this idea is a renewed focus on how to present proofs in computer-friendly ways, which are often totally different from anything that a human would find palatable. Computers don't object if they're told to do the same thing millions of times, or to check two strings of a thousand binary digits to make sure they're identical. They just get on with the job.

Human mathematicians like proofs best when they tell a story, with a clear start, middle, and end, and a compelling storyline that propels you from the start, the hypotheses of the theorem, to its conclusion. Narrative is more important than nitpicking logic. The aims are to be clear, concise, and above all *convincing*. Bear in mind that mathematicians are notoriously difficult to convince.

Computer scientists studying machine-checkable proofs came up with an entirely different approach: interactive proofs. Instead of presenting a proof as a story, written by one mathematician and read by another, this turns the proof into a dispute. One mathematician, traditionally called Pat, wants to convince Vanna that his proof is correct; Vanna wants to convince him it's wrong. They keep asking each other questions and providing answers until one of them concedes. (Pat Sajak and Vanna White were popular American TV personalities on the game show *Wheel of Fortune*.) It's like a game of chess, where Pat claims 'checkmate in four moves'. Vanna disagrees, so Pat makes a move. Vanna counters with 'what if I do *this*?', and Pat makes another move. This to-and-froing continues until Vanna loses. Now she starts to backtrack. 'Suppose my last move had been *this* instead?' Pat makes a different move, checkmate! And so it goes until all Vanna's possible replies to Pat's moves are exhausted, and Pat wins, or until he's forced to admit that, actually, it wasn't checkmate in four moves at all. In my experience, this is exactly what real mathematicians do when they're working together to solve a research problem, and it can get quite heated. The narrative version is how you present the eventual outcome in a seminar.

László Babai and others parlayed this type of argumentative proof technique into the notion of a transparent proof, using mathematical tools such as polynomials over finite fields and error-correcting codes.[31] With these methods established, it was realised that computers can exploit a feature that clarity and conciseness avoid: redundancy. It turns out that a logical proof can be rewritten in a way that makes it vastly longer, but also means that if there's a mistake, it shows up almost everywhere. Every step in the logic is smeared out across the whole proof in multiple related near-copies. It's a bit like a hologram, where an image is transformed so that it can be reconstructed from any small part of the data. You can then check the proof by taking a small random sample. Any mistake will almost surely show up in that sample. Do this, and you've got a transparent proof. The theorem about the nonexistence of class-P approximate solutions is a consequence.

*

Back to that pigeon paper by Gibson, Wilkinson, and Kelly in *Animal Cognition*. They start with the remark that the TSP had recently been used to examine aspects of cognition in humans and animals, especially the ability to plan actions before taking them. However, it wasn't clear whether this ability was restricted to primates. Can other animals also plan ahead, or do they just use rigid rules, developed by evolution? The researchers decided to use pigeons in laboratory trials that presented them with simple TSPs having two or three destinations – feeders. The pigeons start from one location, travel to each feeder in some order, and continue to a final destination. The team concluded that 'Pigeons weighed the proximity of the next location heavily, but appeared to plan ahead multiple steps when the travel costs for inefficient behaviour appeared to increase. The results provide clear and strong evidence that animals other than primates are capable of planning sophisticated travel routes.'

In an interview, the researchers explained the link to the bus-driving pigeon. They suggested that the driver might have had two reasons for objecting: the obvious one of safety, or the worry that the pigeon would be unable to follow a route that would pick up passengers efficiently as the bus drove through the city. As the title of the paper indicates, the team concluded from their experiments that the second worry was unjustified.

Let the pigeon drive the bus.

*

If the world's governments and car manufacturers get their way, then quite soon neither the bus driver nor the pigeon will drive the bus. Instead, the bus will drive the bus. We're heading into the brave new age of self-driving vehicles.

Or maybe not.

The most difficult aspect of self-driving vehicles is to ensure that they interpret their surroundings correctly. Equipping them with their own 'eyes' is easy, because small high-resolution cameras are now manufactured by the billion. But vision needs a brain as well as eyes, so cars, trucks, and buses are being equipped with computer vision software. Then they'll know what they're looking at, and be able to react accordingly.

According to manufacturers, one potential advantage of self-driving vehicles is safety. Human drivers make mistakes and cause accidents. A computer doesn't get distracted, and with enough research and development, a computer driver ought to be safer than any human. Another is that you don't have to pay a bus to drive itself. A big disadvantage, aside from job losses for drivers, is that this technology is still in its infancy, and the systems currently available don't live up to the hype surrounding the technology. A few bystanders and test drivers have been killed already in accidents, yet fully driverless vehicles are now being tested on city streets in several countries. The rationale is that they have to be tested in the real world, and that ultimately they'll save more lives than they take. The alacrity with which regulators have fallen for this seductive line of argument is remarkable. If anyone suggested testing a new drug on random people, without their knowledge or consent, on the grounds that this will save more people than it will

Two images, differing in only a few pixels, shown to the
InceptionV3 network, which classified the left-hand image
as a cat and the right-hand one as guacamole.

kill, there would be an outcry. In fact, it would be illegal in almost
all countries, and definitely unethical.

The principal technology behind computer vision, for this
purpose, is the even more fashionable area of machine learning.
A deep learning network, which adjusts its connection strengths
so that it correctly identifies images, is trained with a huge
number of images until it attains an acceptable level of accuracy.
This procedure has been extremely successful in a wide range of
applications. However, in 2013 it became apparent that too much
attention was being paid to the successes of machine learning,
and too little to potential failures. A serious issue is 'adversarial
examples' – deliberately modified images that a human gets right
but a computer gets spectacularly wrong.

The picture shows two images of a cat. Obvs. They differ in
just a few pixels, and to us they look identical. A standard neural
network, trained on huge numbers of images of cats and not-
cats, correctly identifies the left-hand picture as a cat. It insists,
however, that the right-hand image is guacamole, a green Mexican

sauce made from avocado. In fact, the computer is confident at the 99% level that it's guacamole, compared with only 88% for the cat. As the saying goes, a computer is a device for making millions of mistakes very rapidly.

Images of this kind are said to be 'adversarial' because they arise when someone deliberately tries to fool the system. In practice, the computer will perceive most images like this as a cat. Christian Szegedy and collaborators noticed that such images exist in 2013.[32] In 2018 Adi Shamir and coworkers[33] explained why adversarial examples can occur in deep learning systems, why they are inevitable, and why only a few pixels need to be changed to mislead the neural network.

The root cause of this susceptibility to major errors is dimension. The usual way to measure how different two bit strings are is to find their Hamming distance: how many bits need to be swapped to convert one to the other. So, for instance, the Hamming distance between 10001101001 and 10101001111 is four, the different bits being the four boldface digits in 10**1**0**1**00**11**1**1**. An image is represented in the computer as a very long bit string. If the image occupies 1 MB (megabyte), its length is 2^{23}, or about 8 million bits. So the space of images has dimension 8 million, over the finite field consisting of 0 and 1. It contains $2^{8,388,608}$ different points.

The image recognition algorithm embodied in a trained neural network has to place every image in this space into a far smaller number of categories. In the simplest case, this boils down to dissecting image space into regions by drawing hyperplanes, a procedure illustrated for a two-dimensional space in the picture. This divides space into numerous cells, one for each category. If we change the image to one at a Hamming distance of, say, 40, then we change just 40 bits in the image. The eye is receiving 8 million bits, so this is 0·0005% of the bits, well below the threshold at which a human would notice any significant difference. However, the number of images at this Hamming distance is 2^{50}, about one quadrillion. This is much larger than the number of

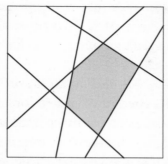

Segmenting image space by hyperplanes. Here the dimension is two, and five hyperplanes (here lines) divide it into 13 cells. One is shown shaded.

categories that the computer vision system can distinguish. So it's not surprising that such a small change to the image can make the computer misread it.

For mathematical analysis it's convenient to represent bit strings not over a finite field but as real numbers. For instance, a single byte of eight bits, say 10001101, can be considered as the real number with binary expansion 0·10001101. Now the space of all 1 MB images becomes a real vector space of dimension one million. With this modification, Shamir and colleagues prove something far stronger. Given an image in one cell of the hyperplane arrangement, and a second cell, how many bits do we need to change in the image to move it into the second cell? Their analysis shows that, for example, if image space is partitioned into a million cells using 20 hyperplanes, then only two coordinates need to be changed to move a given point to any cell whatsoever, provided the dimension of image space is more than 250. In general, if the network has been trained to distinguish a given number of categories, the number of coordinates that must be changed to move a given image into *any* category is about the same as the number of categories.

They tested this theorem on a commercial number-recognition system. Here there are ten categories, the digits 0–9. They generated adversarial images that could persuade the system to recognise the digit 7 as any of the ten possibilities 0–9. Only 11 bits needed to be changed to achieve this. The same goes for any digit other than 7.

Should we worry? 'Natural' images, of the kind our self-driving car will usually encounter, aren't constructed deliberately to fool the system. However, the car will observe around half a million images per day, and it takes only one wrong interpretation to cause an accident. The main threat is that vandals or terrorists can easily modify road signs by adding small pieces of black or white tape, fooling the computer into thinking that a STOP sign is actually a 60 mph speed limit. All of which adds to the feeling that the introduction of self-driving cars is being unduly and unsafely rushed because of commercial pressures. If you disagree, let me repeat: we would *never* introduce a new drug or medical procedure in such a slipshod manner. Especially if there were good reasons to suspect it might be dangerous.

Don't let the bus drive the bus.

4

The Kidneys of Königsberg

In addition to that branch of geometry which is
concerned with magnitudes, there is another branch,
which Leibniz first mentioned, calling it the geometry
of position … Hence, when a problem was recently
mentioned, which seemed geometrical but did not
require the measurement of distance, I had no doubt
that it was concerned with the geometry of position. I
have therefore decided to give here the method that I have
found for solving this kind of problem.

> Leonhard Euler, *Solutio problematis ad geometriam situs*
> *pertinentis*, 1736

Throughout most of human history, the organs you were born
with were also those you died with, and often died *of*. If your
heart failed, or your liver, or your lungs, or your intestines, or
your stomach, or your kidneys, so did you. A few bodily parts,
especially arms and legs, could be surgically removed, and if you
survived the experience, you could lead some kind of life. The
invention of anaesthetics and sterile conditions in operating the-
atres made operations less painful, at least while they were being
performed and the patient was unconscious, and greatly increased
the chances of survival. When antibiotics arrived, infections that
had previously been fatal could often be cured.

We take these miracles of modern medicine for granted, but
they made it possible, essentially for the first time, for doctors and

surgeons to *cure* diseases. We've managed to fritter away most of the advantages of antibiotics by giving them to farm animals on a massive scale; not to treat disease as such, but to make them grow bigger and faster. And by millions of people stopping antibiotic treatment as soon as they felt better, instead of taking the full course of pills like the doctor told them. Both practices, completely unnecessary, have encouraged the development of antibiotic resistance in bacteria. Now scientists are frantically grubbing around trying to find the next generation of antibiotics. If they do, I hope we have the sense not to ruin those too.

Another dream of the surgeons of the past has also been realised: organ transplants. So far we seem to have managed not to wreck that. If circumstances are favourable, you can get a new heart, or a new lung, or a new kidney. Even a new face. One day a kindly pig might even grow a replacement organ for you, though not voluntarily.

In 1907 the American medical researcher Simon Flexner speculated on the future of medicine, suggesting that it would become possible surgically to replace diseased organs by healthy ones from another person. In particular he mentioned arteries, the heart, the stomach, and kidneys. The first kidney transplant was carried out in 1933 by the Ukrainian surgeon Yuriy Vorony, who removed a kidney from a donor who had died six hours previously and implanted it in his patient's thigh. The patient died two days later when the new kidney was rejected because its donor had the wrong blood group. The biggest obstacle to successful organ transplantation is the body's immune system, which recognises the new organ as not being part of the patient's own body, and attacks it. The first successful kidney transplant was performed by Richard Lawler in 1950. The donated kidney lasted ten months before being rejected, but by then the patient's own kidneys had recovered enough for her to live for a further five years.

A normal person has two kidneys, and can function very well with only one of them. So transplants can be obtained from a living donor, simplifying the whole process. The kidney is the easiest organ to transplant. It's straightforward to make sure the donor's tissue type matches the recipient's, preventing rejection, and if something goes wrong, dialysis machines are available to take up the kidney's task. Until anti-rejection drugs came on the scene, which happened in 1964, there were no kidney transplants from deceased donors (at least, in the USA and the UK). But there were plenty donated by living volunteers.

In most cases, the donor was a close relative of the recipient. This increased the likelihood of a tissue match, but the main reason was that few people were willing to sacrifice a kidney for a stranger. After all, when you have a spare you can continue living a normal life if one happens to pack up. If you've given one kidney to a stranger, this backup is lost. If the recipient is your mother, or brother, or daughter, the gain outweighs the risk, especially if you know they'll die if you refuse. It's less personal with a stranger, and you're less likely to take the risk.

Some countries offered an incentive: money. You could pay a stranger to donate a kidney to one of your relatives. The dangers of allowing this kind of transaction are fairly obvious: poor people being bribed into donating a kidney for a rich stranger, for example. In the UK, it was made illegal to donate a kidney to anyone other than a close relative. Laws passed in 2004 and 2006 removed this barrier, but added safeguards to prevent abuses. 'No money to change hands' was one of them.

The change in the law opened up new strategies for matching donors to recipients, making it possible to treat many more patients. It also opened up a significant set of mathematical problems: how to use these strategies efficiently. As it happened, powerful tools to solve these problems already existed. Remarkably, it all started with an absurd little puzzle, nearly 300 years ago.

*

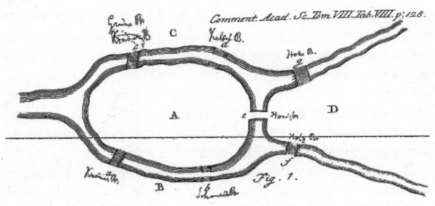

Euler's schematic diagram of the seven bridges of Königsberg.

It's a well-known story, but I'll tell it anyway, for two reasons. It sets up the mathematics, and it's common to get the history wrong. I've certainly done so.

Kaliningrad, which today is in Russia, was once called Königsberg. In the 1700s it was in Prussia. The river Pregel flowed through the city, and there were two islands, Kneiphof and Lomse. There were seven bridges. Each bank of the river was linked to Kneiphof by two bridges; each bank was linked to Lomse by one bridge; finally, a bridge linked the two islands. Today's layout is rather different. The city was bombed in the Second World War and bridges b and d in the picture were wrecked. Bridges a and c were demolished to make way for a new road, and replaced. Together with the remaining three originals, of which one was rebuilt in 1935, there are now five bridges in the original locations.

Legend has it that the good citizens of Königsberg had long wondered whether it was possible to take a walk through the city that crossed each bridge exactly once. It was a simple little puzzle, the kind of thing you'd expect to see nowadays on the puzzle page of your newspaper or its electronic equivalent. Experiments with different paths don't lead to solutions – try it. However, similar problems do have a solution, and sometimes that's hard to find.

Moreover, the number of paths you might take is infinite, if only because there are infinitely many ways to wander from side to side, or back and forth, as you walk along a path. So you can't find a solution, or prove there isn't one, by considering every possible path.

You can solve the puzzle easily by concocting some kind of cheat. For example, you could walk onto a bridge, turn round, and come back off, without actually stepping off at the far end, and claim to have 'crossed' it. The condition 'crossing' must be explicitly defined so that this isn't allowed. Similarly, 'walk' means you can't do part of the journey by swimming, taking a boat, flying in a balloon, or hitching a ride in Doctor Who's TARDIS. Or by wandering off upriver to find a bridge that's not in Euler's picture. Puzzle enthusiasts know that while 'cooking' the puzzle like this may be fun, and may even require great ingenuity, it's a cheat. I'm not going to state every single condition needed to rule out this kind of cookery. I'm much more interested in how the puzzle, suitably reformulated as mathematics, can be proved impossible *unless* you cook it. Cookery is about how the problem could be formulated, not about solving it or proving it impossible once it's *been* formulated.

Enter Euler, the leading mathematician of his age. He worked on just about every area of mathematics there was, and some that weren't until he started them going, and he applied the subject to a huge variety of real-world problems. His work ranges from erudite tomes on major areas of pure mathematics and mathematical physics to curiosities and oddities that happened to catch his fancy. In the early 1700s he turned his mind to the puzzle of the Königsberg bridges. He formulated it as a precise mathematical question, and came up with a proof that, as stated, no such walk can exist. Not even if it's not a round trip, but ends somewhere different from where it starts.

Euler had moved to St Petersburg in Russia in 1727, when Russia was ruled by the Empress Catherine I, to become court mathematician. Her husband, the Emperor Peter I, had founded the St Petersburg Academy (Academia Scientiarum Imperialis Petropolitinae) in 1724–5, but died before it had fully come into being. Euler presented his work to the Academy in 1735, and it was published a year later. Being a mathematician, arguably the most prolific in history, he extracted as much from the puzzle as he could: he found necessary and sufficient conditions for a solution to exist, not just for the Königsberg bridges but for any problem of a similar kind. You can have fifty thousand bridges linking forty thousand landmasses in some hugely complex arrangement, and Euler's theorem still tells you whether a solution exists. If you look hard at the proof, it even tells you how to find one – after a bit of messing around. Euler's discussion was a bit sketchy and it took nearly 150 years before anyone sorted out all the details, although it's not terribly difficult.

At this point many books on graph theory tell you that Euler proved the puzzle has no solution by reducing it to an apparently simpler question about *graphs*. A graph, in this sense, is a set of dots (called nodes or vertices) joined by lines (called edges), forming a kind of network.[34] The reformulation using graphs converts the problem of the Königsberg bridges into that of tracing a path through a particular graph that uses each edge exactly once. That's certainly how we tackle the question today, but it's not quite what Euler did. History is like that. Historians of mathematics delight in telling you what actually happened, instead of what the standard story says. Actually, Euler solved the whole shooting match symbolically.[35]

He labelled each region of land (island or river bank) and each bridge with a letter. He used capital letters A, B, C, D for the land, and lower case letters a, b, c, d, e, f, g for the bridges. Each bridge joins two different regions, for instance bridge f joins A to D. A walk starts in some region, and can be specified by listing, in order, which regions it meets and which bridges it crosses, ending with the final region visited. For much of his paper Euler does this verbally, and mostly he just worked with the sequence of regions. It doesn't matter which bridge you use to go from A to B; just that the number of occurrences of AB is the same as the number of such bridges. Alternatively, you can just use the sequence of bridges, provided you specify where to start, and count how many times you encounter a given region. This would arguably have been simpler. Towards the end of the paper he uses both symbols, giving an example using the sequence

EaFbBcFdAeFfCgAhCiDkAmEnApBoElD

corresponding to a more complicated layout.[36]

In this formulation, the precise path that the walker follows, in each region or along each bridge, is irrelevant. The only thing you need to keep track of is the sequence in which regions are visited and bridges are crossed. Crossing a bridge is interpreted as 'the two capital letters on either side are different'. This rules out wandering onto a bridge and off again at the same end. A solution is a sequence of alternating capital and lower case letters A–D and a–g in which each lower case letter appears once and only once, and in which the capital letters before and after a given lower case one correspond to the two regions it links.

We can list these links, for every lower case letter:

a	joins	A and B
b	joins	A and B
c	joins	A and C
d	joins	A and C
e	joins	A and D
f	joins	B and D
g	joins	C and D

Suppose we start from region B. Three bridges link B to some other region: a, b, or f. Suppose we choose f; then the sequence starts Bf. The region at the other end of f is D, so now we have BfD. Two bridges linking D to another region remain unused: e and g. (We can't use f again.) Let's try g, so now the walk is BfDg. The other end of g is C, giving BfDgC. Now bridges c and d are the only ways to continue (we can't go back over g). Maybe we try bridge c, which leads to BfDgCc and then BfDgCcA. From region A there are four possible bridges: a, b, d, and e. (We've used up c.)

Can we now cross d? No, because that gives BfDgCcAd and then BfDgCcAdC. Now all three bridges linked to C have been used up – namely, c, d, and g. But we haven't solved the puzzle, because bridge b hasn't been crossed. Scrub bridge d. For similar reasons we can't exit across bridge e: that takes us to D, and we're stuck; moreover, we've missed out b again. What about a? This gives BfDgCcAaB, and the only unused exit is via b, giving BfDgCcAaBbA. The only exits now are d or e. The first leads to BfDgCcAaBbAdC, with no exit left – but we haven't crossed e. The second leads to BfDgCcAaBbAeD, with no exit left – but we haven't crossed d.

OK, so that series of choices doesn't work, but we could have made different choices earlier on. We can now work systematically through all possible sequences ... and it turns out that all of them can be eliminated. At some stage you get stuck, with no exit from your current region, but with at least one bridge still not crossed. The list of possible sequences is finite, and small enough to write down completely. Try it if you wish.

Assuming you've done that, you've proved this particular puzzle has no solution. That might have satisfied the citizens of Königsberg, but it didn't satisfy Euler. First, it's not clear *why* you always get stuck. Second, the answer doesn't tell you when other puzzles of the same kind can or can't be solved. So Euler asked the single most important question mathematicians always ask when someone solves a problem: 'Yes, but why did that work?' Followed by the next most important question: 'Can we do better?'

Euler thought some more, and made three simple observations:

- If there's a solution, every region must be connected to every other one by *some* sequence of bridges. For example, if there were two more islands E and F joined to each other by one or more new bridges h, i, j, …, with no other new bridges linking those islands to the other regions, then the only way to cross those bridges is to shuttle back and forth between E and F. So you can't get to any of the other bridges.
- Assuming the previous 'connectedness' condition is OK, then except for the two regions at the start and end of your walk, whenever you enter a region you must exit it again, over a different bridge.
- Whenever you do that, two bridges linked to that region cease to be available.

Therefore, as you walk along, you use up bridges in pairs. This is the key insight. If a region sits on the end of an even number of bridges, you can use them all up without getting stuck in that region. If it sits at the end of an odd number of bridges, you can use up all but one of them without getting stuck. But *you have to cross that bridge* at some stage. And when you do, you're stuck.

Getting stuck is fatal if you're in the middle of a hypothetical tour. However, it's not an issue if you're at the end of the tour. Reversing the tour and, in effect, walking backwards, it's also not an issue at the start of the tour. This line of reasoning implies that if a tour exists, at most two regions must lie at the end of an odd number of bridges. For the Königsberg problem:

A   is linked by 5 bridges
B   is linked by 3 bridges
C   is linked by 3 bridges
D   is linked by 3 bridges

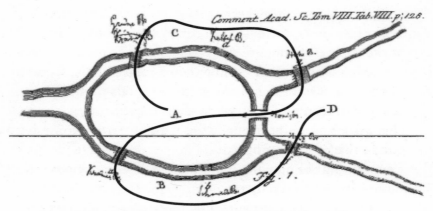

Open-ended tour using the five bridges still remaining.

Therefore the number of regions at the end of an odd number of bridges is four, which is more than two. So no tour exists.

Euler also stated, without proof, that the same odd/even condition is sufficient for a tour to exist. This is a bit harder and I won't go through it; it was proved by Carl Hierholzer just before his death in 1871, and published posthumously in 1873. Euler also remarked that if you're looking for a closed tour, ending where it started, then a necessary and sufficient condition is that every region lies at the end of an even number of bridges.[37]

Using only the five bridges that (in some form) survive today, both B and C are linked by two bridges. Therefore this revised problem must have a solution, but only for a closed tour. The end points must be on A and D since those are still linked to an odd number of bridges. The picture shows such a solution. There are others: can you find them all?

Euler phrased all of the above in terms of symbolic sequences like BfDgCcAaBbAeD. Some time later, somebody realised that you can give everything a visual interpretation. Precisely who isn't clear, because the idea was very much in the air around the mid-nineteenth century, but James Joseph Sylvester introduced the name 'graph' in 1878. Draw a picture with four dots A–D, and seven lines a–f. Make each line join the two regions at the end of the corresponding bridge. The map of islands and bridges now simplifies, as in the left-hand picture. The symbolic sequence just mentioned corresponds to the path in the right-hand picture, starting at B and ending at D, where it gets stuck.

This visual simplification is the *graph* of the Königsberg bridges. In this representation, it doesn't matter where you place the four dots (although they should remain distinct to avoid getting confused between them) and the precise shapes of the lines also don't matter. All that matters is which dots a given line connects. In this visual setting, Euler's proof becomes very natural. Any tour that enters a region across one bridge must leave it again across another, unless it's the end of a closed tour. Similarly, any tour that leaves a region across one bridge must have entered it already across another, unless it's the start of a closed tour. So bridges come in pairs except at the two ends. Therefore, the regions not at the ends meet an even number of bridges. If the ends meet an odd number, only an open tour is possible. Alternatively, the start and end might be in the same region, so you can join them together without using any more bridges, creating a closed tour. Now every region meets an even number of bridges.

In solving this single class of problems, Euler managed to kick-start two major areas of mathematics. One is graph theory, which studies dots joined by lines. It sounds simple, even childish. It is. At the same time, it's profound, useful, and difficult, as we'll see. The other is topology, sometimes called 'rubber sheet geometry', in which shapes can be deformed continuously without being considered essentially different. Here the shapes of the lines and locations of the dots can be deformed any way you wish, provided the way they connect doesn't change (the continuity requirement), and you get essentially the same graph. The same, in the sense that it conveys the same information about what connects to what.

I find it remarkable that such a simple puzzle can lead to

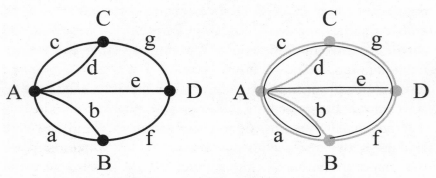

Left: Graph showing connections for Königsberg bridges.
Right: Sample attempt at a tour – bridge d is omitted.

such significant innovations. Unreasonable effectiveness, indeed. There's also an important lesson, which the outside world often fails to grasp. Don't underestimate mathematics that *looks* simple, more like a child's toy than anything serious. What counts is not how simple the toy is; it's what you do with it. Indeed, a primary aim of good mathematics is to make everything as simple as possible. (You may laugh, which is fair enough given how complicated much of it looks. I must add the caveat attributed to Einstein: as simple as possible, *but not more so*.) Reducing islands to dots and bridges to lines doesn't change the puzzle, but it does strip away extraneous information – what's the weather like? Is the ground muddy? Is it a wooden bridge or a metal one? These things matter, if you're going out for a Sunday stroll or building a bridge. But if you want to answer the question that puzzled the good citizens of Königsberg, they're extraneous clutter.

*

What do the Königsberg bridges have to do with kidney transplants? Directly, not a lot. Indirectly, Euler's paper pioneered the development of graph theory, which opens up a powerful way to match donors to recipients even when most donors only want to donate their kidney to a close relative.[38] When the UK's Human Tissue Act came into force in 2004, people could legally donate kidneys to non-relatives.

A major problem is matching donors to recipients, because even when there's a willing donor, their tissue and blood types may not match those of the intended recipient. Suppose Uncle Fred needs a kidney, and his son William is willing to donate one, though not to a stranger. Unfortunately, William's kidney has the wrong tissue type. Before 2004, that was the end of the story, and Fred was stuck with frequent treatment on a dialysis machine. So were many other potential recipients whose tissue types matched William's. Now suppose that John Smith, unrelated to Fred and William, has the same problem: his sister Emily needs a new kidney, and he's willing to donate one to her, but, again, not to a stranger. And once more his tissue type differs from Emily's. So no one can get a transplant.

However, suppose that John's tissue type matches Fred's, and William's tissue type matches Emily's. After 2004, this sets up conditions for a legal kidney swap. The surgeons concerned can get together, and suggest that John allows his kidney to go to Fred, on condition that William's kidney goes to Emily. Both donors are much more likely to agree to such an arrangement, because their relative gets a new kidney, while each of them makes a donation that they were always willing to make provided it was for the good of their relative. Exactly who gets which kidney makes little difference either to donors or recipients, although it's crucial to matching tissue types.

With modern communications, surgeons can find out when this kind of coincidence occurs by keeping a register of potential donors and recipients, along with their tissue types. When the number of recipients and potential donors is small, this convenient type of swap is unlikely, but it becomes much more likely when the numbers get bigger. The number of potential recipients is quite large: in 2017 in the UK more than five thousand people were on the waiting list for a new kidney. It could come from a deceased donor or a living one, but the number of donors is smaller – around two thousand at that time, leading to a typical waiting time of over two years for an adult and nine months for a child.

One way to ensure that even more patients benefit, and get treated more quickly, is to set up more elaborate chains of kidney swaps. The law now allows this set-up as well. Suppose that Amelia, Bernard, Carol, and Deirdre all need kidneys. They all have donors lined up, who initially are willing to donate to them – but only them. Suppose these are Albert, Beryl, Charlie, and Diana. The chain starts with an altruistic donor Zoe, who is happy to donate a kidney to anyone. Suppose that the tissue types permit a chain of the following kind:

> Zoe donates to Amelia.
> Amelia's donor Albert agrees to donate to Bernard.
> Bernard's donor Beryl agrees to donate to Carol.
> Carol's donor Charlie agrees to donate to Deirdre.
> Deirdre's donor Diana agrees to donate to the waiting list.

On the whole, everyone is satisfied. Amelia, Bernard, Carol, and Deirdre get new kidneys. Albert, Beryl, Charlie, and Diana all donate one of theirs – not to their relative, but as part of a chain that *benefits* their relative. Often they will be happy with this, which is what makes such transactions possible; in fact, if the don't agree, their relative won't get a kidney on this occasion. Zoe is happy that her altruistic donation benefits someone, and doesn't care whom. In this case, it's Amelia. Finally, an extra kidney goes onto the waiting list – always useful.

If instead Zoe had donated to the waiting list, the only way for Amelia, Bernard, Carol, and Deirdre to get kidneys would be to join the waiting list. By not doing that, they free up four more kidneys. It's called a domino-paired donation chain. Zoe topples a domino, and a whole series of them topple in turn. Let's abbreviate that to *chain*.

What matters here is not the names, but the tissue types. Albert is anyone with the same tissue type as Zoe. Betty is anyone with the same tissue type as Albert's *donor*, Charlie is anyone with the same tissue type as Betty's donor, and so on. With reasonable numbers of recipients and donors, such chains are common, and the surgeons can spot them. However, this takes time, even if they delegate the task, and every kidney is precious, so it makes sense to choose the chains in the best possible way. This is complicated, because many potential chains can coexist. If so, the surgeons can get on with the job simultaneously, unless two chains contain the same donor, but require them to donate to two different people. Then one chain is broken.

Optimise the choice of chains ... Hmm. Sounds mathematical. If you can formulate the problem in mathematical terms, and apply suitable techniques, maybe you can solve it. Moreover, the solution need not be perfect. Just better than anything you could achieve by guesswork. David Manlove found a way to turn the kidney-swap problem into a question about a graph. Euler's theorem doesn't help solve it – his role was to found the entire area. Over the intervening years, mathematicians have developed the subject and invented many new graph-theoretic techniques. Because a graph is a discrete object, 'really' just a list of nodes, edges, and which edges join which nodes, graphs are eminently suited to manipulation by computer. Powerful algorithms have been developed to analyse graphs and extract useful structure. Among them are algorithms that can find optimal allocations of donors to patients, for graphs of practical size. These methods, implemented by computer, are now in routine use in the UK.

*

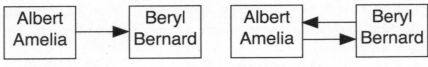

Two types of swap.

Compatible pairs of donors and recipients are easy: swap their kidneys. This requires two surgeons to operate at the same time, one on each person. So we can ignore compatible pairs when trying to find chains, and focus on incompatible pairs. These *pairs* constitute the nodes of the graph.

For instance, suppose that Albert is willing to donate to Amelia, but compatible with Bernard. We can represent this situation by the left-hand picture. I put the donor's name on top and their incompatible relative below. The arrow means 'donor at the tail end is compatible with recipient at the head end'. This picture is a special type of graph, in which the edges have specific directions. Unlike the bridges of Königsberg, these edges are one-way: mathematicians call them directed edges, and the resulting graph is a directed graph, or digraph for short. In a drawing, directed edges are represented by arrows.

If it so happens that Beryl is compatible with Amelia, the rules tell us to draw another arrow in the opposite direction. This creates a two-way connection, as in the right-hand picture. This picture illustrates the simplest kind of kidney swap, which graph-theorists call a 2-cycle. The surgeons can suggest that Albert donates his kidney to Bernard, on condition that Beryl donates hers to Amelia. If all parties agree, then Amelia and Bernard get new kidneys, while Albert and Beryl donate one kidney each. Although their relative doesn't get *their* kidney, they do get *a* kidney. Both recipients benefit, and both donors donate, so most potential donors are willing to accept this kind of swap.

The next more complicated kind is a 3-cycle. Now there's a third pair, with donor Charlie and recipient Carol. Suppose that

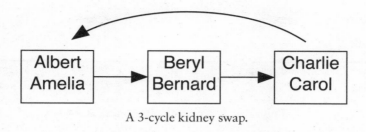

A 3-cycle kidney swap.

> Albert is compatible with Bernard
> Beryl is compatible with Carol
> Charlie is compatible with Amelia

Then the surgeons can arrange to give Albert's kidney to Bernard, Beryl's to Carol, and Charlie's to Amelia. Again, most donors would accept this.

Altruistic donors like Zoe have to be dealt with slightly differently, because they're not paired with any specific person. A little bit of mathematical trickery now comes into play. We form the corresponding node by pairing Zoe with a recipient labelled 'anyone', deemed to be compatible with any of the non-altruistic donors. In practice, this dummy recipient represents all of the people on the waiting list. The assumption is that each of them is compatible with *some* non-altruistic donor, which is reasonable because the waiting list is big. Now we draw an arrow from

$$\text{node-Z} = (\text{Zoe, anyone})$$

to any node whose recipient is compatible with Zoe. In the domino-paired chain described earlier, the digraph looks like the next picture.

Chains like this aren't practical: this one would require ten surgeons to operate simultaneously. It has to be essentially simultaneous, otherwise – say – Charlie could suddenly change his

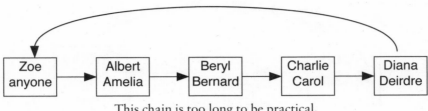

This chain is too long to be practical.

mind and refuse to donate to Deirdre, once Carol has received a kidney from Beryl. People don't always stick to agreements when it becomes to their advantage not to, even if they've signed legal documents. If they want to, they'll find some way to weasel out. Plead illness, whatever. Break a leg.

For this reason, swaps are currently limited to four scenarios: 2-cycles and 3-cycles, which we've already seen, and the corresponding cycles when an altruistic donor is involved, which are called short chains and long chains. A short chain would involve only Zoe, Albert, Amelia, and someone on the waiting list. A long chain would also include Beryl and Bernard. An *exchange* is any of these four scenarios.

Notice the slight trickery. I've shown this chain as a cycle with five nodes. When realised as a swap, it's not quite that, because Zoe has no specific recipient in mind. Zoe is willing to donate to anyone, and at the end of the chain Diana *does* donate to anyone (that is, to the waiting list). But the anyone that Zoe actually donates to is Amelia, and that's not the same person that Diana ends up donating to. The mathematics takes care of this, because we infer who 'anyone' is in each case from the structure of the digraph.

The digraphs above show individual cycles and chains, using small numbers of nodes and arrows. In reality, there are a lot of pairs, a rather smaller number of altruistic donors, and a huge number of arrows. This happens because the digraph must have an arrow between *any* two nodes X and Y such that the donor of

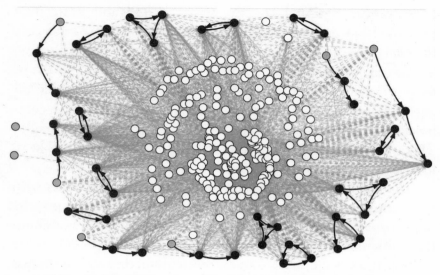

Kidney exchange digraph used in July 2015, and optimal solution (solid black lines). White dots = unmatched donors and recipients, grey dots = altruistic donors, black dots = matched recipients/donors.

X is compatible with the recipient of Y. The same donor can be compatible with many recipients. For example, in October 2017 a total of 266 non-altruistic pairs and 9 altruistic donors had been registered, and there were 5,964 arrows. The picture illustrates similar complexity on another date. The mathematical challenge is to find not just one exchange in the digraph, but the best possible set of exchanges.

*

In order to solve this problem mathematically, we need to make precise the notion of 'best possible'. It's not just the set of exchanges that involves the largest number of people. There are other considerations, such as cost and likely success rate. Here, medical advice and experience comes into play. The UK organisation National Health Service Blood and Transplant (NHSBT) has developed a standard scoring system to quantify the likely benefit of any particular transplant. This takes account of such factors as how long a patient has been waiting, levels of tissue-type incompatibility, and the age difference between donor and recipient. Using a statistical analysis, these factors are combined in a single numerical score, a real number called the weight. This weight is calculated for every arrow in the digraph – that is, for every potential transplant.

One condition is straightforward: two different exchanges in the set can't involve any common nodes, because you can't give the same kidney to two different people. Mathematically, this states that the component cycles in the set don't overlap. The others are subtler. Some 3-cycles have an extra useful feature: an additional arrow in the reverse direction between two nodes. The picture shows a case where, in addition to the arrows in the previous 3-cycle, there's another one from Beryl's pair to Amelia's. That is, Beryl is compatible with Amelia as well as with Carol. If at a late stage Charlie drops out of the swap arrangement, Carol can then be removed as well; there remains a 2-cycle in which Albert donates to Bernard and Beryl donates to Amelia, and that swap could still go ahead. Mathematically, this trio of nodes forms a 3-cycle, and in addition two of the nodes form a 2-cycle. The extra arrow is called a back-arc. Any 3-cycle with a back-arc, and any 2-cycle, is called an *effective* 2-cycle.

The Kidney Advisory Group of NHSBT defined a set of kidney exchanges to be optimal if:

(1) It maximises the number of effective 2-cycles.
(2) It contains as many cycles as possible, subject to (1).
(3) It uses as few 3-cycles as possible, subject to (1) and (2).
(4) It maximises the number of back-arcs, subject to (1)–(3).
(5) It maximises the total weight of its cycles, subject to (1)–(4).

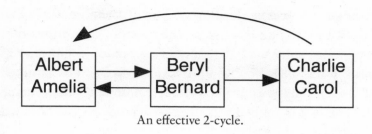

An effective 2-cycle.

The intuition behind this definition is that certain features are prioritised; once that's been done, other features with lower priority successively come into play. For example, condition (1) makes sure that including three-way exchanges doesn't reduce the number of two-way exchanges that could otherwise have been made. This has advantages, one being simplicity, another being the possibility of going ahead with the 2-cycle if someone drops out. Condition (5) means that after the main decisions (1)–(4) are taken, and only then, the set of exchanges should be as efficient, and as likely to succeed, as possible.

The mathematical problem is to find an optimal set of exchanges, according to these criteria. A little thought and a few back-of-the-envelope calculations show that it's not feasible to check every possible set of exchanges. There are just too many possibilities. Say there are 250 nodes and 5,000 edges. On average, each node meets 20 edges, and for a ballpark figure we can assume it's at the head of 10 and the tail of 10. Suppose we want to list all possible 2-cycles. Pick a node and follow the 10 arrows emanating from it. Each arrow ends at a different node, each with its own 10 outgoing arrows. We get a 2-cycle if the final node is the same as the first one. That's 100 possibilities to check. Finding a 3-cycle takes 100 × 10 = 1,000 checks of this kind – 1,100 checks per node. With 250 nodes, that's now 275,000 cases to check, ignoring shortcuts that can reduce the total somewhat but don't change the general order of magnitude.

However, all you've done now is to list the possible 2- and 3-cycles. An exchange is a *set* of them, and the number of sets grows exponentially with the number of cycles. In October 2017 the digraph had 381 2-cycles and 3,815 3-cycles. The number of sets of 2-cycles alone is 2^{381}, a 115-digit number. The number of sets of 3-cycles has 1,149 digits. And we haven't yet checked which sets don't have any overlaps.

Needless to say, this is not how you solve the problem. But it does make it clear that some pretty powerful methods have to be invented to do that. I'll just sketch some of the ideas involved. You can think of this problem as a glorified kind of Travelling Salesperson Problem: it's a combinatorial optimisation problem with rather different constraints, and some considerations are similar. The crucial one is how long it takes to calculate an optimal solution. We can examine this strategy from the viewpoint of computational complexity, as in Chapter 3.

If the swaps involve only 2-cycles, an optimal set of exchanges can be calculated in polynomial time, complexity class P, using standard methods for maximum weight matching in a graph. When there are 3-cycles, even without altruistic donors, the optimisation problem is NP-hard. Nonetheless, Manlove and coworkers devised a workable algorithm based on linear programming, which we met in Chapter 3. Their algorithm, UKLKSS, recasts the optimisation problem so that it can be solved using a sequence of linear programming calculations. The result of each is fed into the next one as an additional constraint. So first condition (1) is optimised; this uses a method called Edmonds' Algorithm, as implemented by Silvio Micali and Vijay Vazirani. Edmonds' Algorithm finds the maximum matching in a graph, in a time proportional to the number of edges times the square root of the number of nodes. A matching associates pairs of vertices at the ends of a common edge, and the problem is to match as many pairs of nodes as possible without using two edges that meet at a common node.

Having optimised for condition (1), that solution is fed into the calculation for condition (2), using an algorithm called the COIN-Cbc integer programming solver, part of the Computational Infrastructure for Operations Research project's collection of algorithms, and so on.

By the end of 2017 these graph-theoretic methods had identified a total of 1,278 potential transplants, but only 760 were carried out because all sorts of practical issues can show up in late stages of assessment: the discovery that tissue types aren't as compatible as had been thought, or donors or recipients being too ill to undergo the operation. However, the systematic use of graph-theoretic algorithms to organise kidney transplants efficiently is a big improvement over previous methods. It also points the way to future improvements, because it's now possible to keep kidneys healthy outside the body for longer periods, so the operations in a chain need not all happen on the same day. This makes it feasible to consider longer chains. Raising new mathematical problems.

I'm not trying to give Euler credit for crystal-gazing. He hadn't the faintest idea that his clever solution of a silly puzzle would ever be useful in medicine. Certainly not in organ transplants, at a time when surgery was a painful form of butchery. But I do want to give him credit for seeing, even in those early days, that the puzzle hinted at something much, much deeper. He said so, explicitly. Look at the epigraph at the start of this chapter. Euler repeatedly mentions 'geometry of position' as the relevant context. The actual Latin phrase he used was *analysis situs*. He credits Leibniz with this term, and by implication, with realising that such a subject might be important. He's obviously intrigued by the idea of a form of geometry that's not about traditional Euclidean shapes. He doesn't reject it because it's unorthodox, quite the reverse. He's not hidebound by tradition. And he's pleased to add his own small contribution to the development of such a geometry. He's having fun.

Leibniz's dream came to fruition in the twentieth century,

with some significant advances in the nineteenth. We now call it *topology*, and I'll show you some of its new uses in Chapter 13. Graph theory still has associations with topology, but it has mainly developed along separate lines. Notions such as the weight of an edge are numerical, not topological. But the idea that you can use graphs to model complicated interacting systems, and to solve optimisation problems, goes right back to Euler, tackling a new kind of question *because it caught his imagination*, and inventing his own ways to do that. In St Petersburg, Russia, under Empress Catherine I, nearly three centuries ago. Everyone who gets a transplanted kidney, in the UK or any other country that uses graph-theoretic techniques to allocate organs more effectively, should be delighted that he did.

5

Stay Safe in Cyberspace

No one has yet discovered any warlike purpose to be
served by the theory of numbers or relativity, and it
seems unlikely that anyone will do so for many years.
Godfrey Harold Hardy, *A Mathematician's Apology*, 1940

Pierre de Fermat is famous for his 'Last Theorem', that if n is at
least three, the sum of two nth powers of whole numbers can't
also be an nth power. Andrew Wiles eventually found a modern,
technical proof in 1995, some 358 years after Fermat stated his
conjecture.[39] Fermat was a lawyer at the Parliament of Toulouse,
but spent most of his days doing mathematics. He had a friend
called Frénicle de Bessy, a Parisian mathematician best known for
cataloguing all 880 magic squares of order four. The two corresponded frequently, and on 18 October 1640, Fermat wrote to de
Bessy, telling him (in French) that 'every prime number divides ...
one of the powers minus one of any progression, and the exponent of this power divides the given prime minus one'.

In algebraic language, Fermat was saying that if p is prime and
a is any number, then $a^{p-1} - 1$ is divisible by p (without remainder).
So, for example, since 17 is prime, he's asserting that all of the
numbers

$$1^{16} - 1 \quad 2^{16} - 1 \quad 3^{16} - 1 \quad ... \quad 16^{16} - 1 \quad 18^{16} - 1 \quad ...$$

are exact multiples of 17. Obviously we have to miss out $17^{16} - 1$, which can't be a multiple of 17 because it's 1 less than such a multiple, namely 17^{16}. Fermat knew that this extra condition is needed, but didn't say so in the letter. Let's check one case:

$$16^{16} - 1 = 18,446,744,073,709,551,615$$

and this number divided by 17 equals

$$1,085,102,592,571,150,095$$

exactly. How about that?

This curious fact is now called Fermat's Little Theorem, to distinguish it from his Last (or Great) Theorem. Fermat was one of the pioneers of number theory, which studies deep properties of whole numbers. In his day, and for three centuries afterwards, number theory was the purest of pure mathematics. It had no important applications, and it never looked like getting any. Godfrey Harold Hardy, one of Britain's leading pure mathematicians, certainly felt that way, and said so in his little gem *A Mathematician's Apology*, published in 1940. One of his favourite areas of mathematics was number theory, and together with Edward Maitland Wright he published a classic text *An Introduction to the Theory of Numbers* in 1938. Fermat's Little Theorem is in there, as Theorem 71 of Chapter VI. In fact, the whole chapter is about its consequences.

Hardy's political and mathematical views were coloured by prevailing attitudes at the highest levels of academe, and come across now as rather affected, but his writing is elegant and the insight he affords into the academic attitudes of the time is valuable. Some are even relevant today, but others have been rather overtaken by events. Hardy wrote that: 'It is a melancholy experience for a professional mathematician to find himself writing about mathematics. The function of a mathematician is to do something, to prove new theorems, to add to mathematics, and

not to talk about what he or other mathematicians have done.'
So much for 'outreach', highly valued in today's academic world
– but the same snobbish attitude to communication was still prevalent there forty years ago.

One reason Hardy felt the need to justify his profession was
that, in his view, the kind of mathematics he had devoted his life
to didn't have any useful applications and was unlikely to acquire
any. It wasn't paying its way. His interest in the subject was purely
intellectual: the satisfaction of solving difficult problems, and the
advancement of abstract human knowledge. He wasn't greatly
concerned about utility, but he did feel slightly guilty about that.
What did concern him, as a lifelong pacifist, was that mathematics
should not be used in warfare. The Second World War was raging,
and throughout the ages, some areas of mathematics have had
serious military uses. Archimedes is said to have used his knowledge of the parabola to focus sunlight on enemy ships, setting them
on fire, and the law of the lever to design a huge claw that could lift
them out of the water. Ballistics tells you how to target ordnance,
from cannonballs to explosive shells. Missiles and drones rely on
sophisticated mathematics, such as control theory. But Hardy was
sure that his beloved number theory would never have military uses
– at least, not for a very long time – and was proud of it.

*

Hardy wrote at a time when a Cambridge 'don' (fellow) would
spend about four hours of the day doing research, with maybe the
odd hour of teaching, and the remainder relaxing to recharge the
intellectual batteries. He watched cricket and read the newspaper.
It presumably didn't occur to him that even a leading research
mathematician could also use that spare time to inform non-
specialists about what mathematicians were doing. That way, they
could create new mathematics *and* write about it. Which is what
many of us in the profession do today.

Hardy's general point, that a large amount of 'pure' mathematics has no direct use, and probably never will, is often valid.[40] But, somewhat predictably, as soon as he proposed specific examples of useless topics, he ran the risk of picking precisely the wrong ones. When he said that number theory and relativity were unlikely to serve any warlike purpose for many years, he was dead wrong – though we should give him credit for not ruling such applications out entirely. The big problem is to decide, ahead of time, which ideas will acquire applications, and which won't. Crack that one, and you can make a fortune. But it's exactly the areas that *don't* appear applicable that can suddenly propel themselves to the forefront of industry, commerce, and, unfortunately, warfare, and so it was with number theory. Specifically, with Fermat's Little Theorem, now the basis of what we believe to be unbreakable codes.

The irony is that two years *before* Hardy made his apology, the head of MI6 had purchased Bletchley Park, which would house the Government Code and Cypher School (GC&CS), the clandestine centre for Allied codebreaking during the Second World War. Here teams of cryptanalysts famously broke the Enigma code used by the Germans in the war, along with several other Axis code systems. Bletchley Park's best-known member, Alan Turing, began training in 1938 and arrived there on the day war was declared. The cryptanalysts at Bletchley Park used ingenuity and mathematics to break German codes, and ideas from number theory featured among their methods. Within forty years, a revolution in cryptography, firmly based on number theory, was under way, with important military applications as well as civilian ones. Soon it became vital to the functioning of the Internet. We depend on it today, largely without realising it exists.

Relativity, too, acquired military and civilian uses. It had a peripheral affect on the Manhattan Project to develop an atomic bomb, embodied in the popular myth that Einstein's famous equation $E = mc^2$ convinced physicists that small amounts of matter contain huge quantities of energy. This was mainly a rationalisation used after the attacks on Hiroshima and Nagasaki to give the public an easy way to grasp how such weapons were possible. It may even have been intended to deflect public attention away from the real secret: understanding the physics of nuclear reactions. More recently, and more to the point, the accuracy of the Global Positioning System of satellite navigation (Chapter 11) relies on both special and general relativity to calculate locations correctly, and that was funded by the US military and initially reserved for their use.

Military two, Hardy nil.

I'm not blaming Hardy. He had no idea what was going on at Bletchley Park, and he could scarcely have anticipated the rapid rise of digital computation and communications. 'Digital' basically means working with whole numbers, and that's what number theory is all about. Suddenly, results obtained by generations of pure mathematicians out of intellectual curiosity could be mined for innovative technology. Today, vast quantities of mathematics – not just number theory, but everything from combinatorics to abstract algebra to functional analysis – are embodied in the electronic devices that a quarter of the human race carries around daily. The secrecy of online transactions, by individuals, corporations, and military security services, is ensured by cunning mathematical transformations rooted in Hardy's beloved number theory. This wouldn't have surprised Turing, who was so far ahead of the game that he was thinking seriously about artificial intelligence in 1950. But Turing was a visionary. Back then, it wasn't even science fiction. Just fantasy.

*

A code, or cipher, is a method for converting a message in ordinary language, the plaintext, into a ciphertext that looks like gibberish. The conversion generally relies on a key – a vital item of information that's kept secret. For example, Julius Caesar is said to have used a cipher in which each letter of the alphabet is moved along three places. The key here is 'three'. This type of substitution cipher, in which each letter of the alphabet is transformed into some other letter in a fixed manner, can easily be broken, given an adequate supply of ciphertexts. You just need to know the frequencies with which the letters of the alphabet occur in plaintext. Then you can make a pretty good stab at guessing the code. At first there will be a few errors, but if a section seems to decode as JULFUS CAESAR, you don't need to be a genius to realise that the F ought to be an I.

Simple and insecure though it may be, the Caesar cipher is a good example of a general principle that, until recently, underlay virtually all code systems: it's a symmetric cipher, meaning that both the sender and the recipient use essentially the same key. I say 'essentially' because they use it in different ways: Julius moves the alphabet three places forward, while the recipient moves it three places back. However, if you know how the key is used to encode a message, you can easily reverse the process to use the same key to decode it. Even very sophisticated and secure ciphers are symmetric. So security demands that *the* key must be kept secret, from everyone except the sender and the recipient.

As Benjamin Franklin said, 'three may keep a secret, if two of them are dead'. In a symmetric cipher, at least two people need to know the key, which in Franklin's view is one too many. Some time in 1944 or 1945, someone (perhaps Claude Shannon, inventor of information theory) at Bell Labs in the USA suggested protecting voice communication from eavesdroppers by adding random noise to the signal, and then subtracting it again when received. This is also a symmetric method, because the key is the random noise, and subtraction reverses addition. In 1970 James Ellis, an engineer at the UK's Government Communications Headquarters (GCHQ), formerly GC&CS, wondered whether the noise could be generated mathematically. If so, it was at least conceivable that this could be done not by mere addition of signals, but by some mathematical process that would be very difficult to reverse, even if you knew what it was. Of course, the recipient had to be able to reverse it, but that might be achieved using a *second* key, known only to the recipient.

Ellis called this idea 'non-secret encryption'. Today's term is 'public key cryptosystem'. These phrases mean that the rule for putting a message into code can be revealed to the general public, but without knowledge of the second key, no one would be able to work out how to reverse that procedure and decode the message. The only problem was that Ellis couldn't devise a suitable encryption method. He wanted what's now called a trapdoor function: easy to calculate, but hard to reverse, like falling through a trapdoor. But, as always, there had to be some secret second key that let the legitimate recipient reverse the process just as easily, like a hidden ladder that you could use to climb out again.

Enter Clifford Cocks, a British mathematician also at GCHQ. In September of 1973, Cocks had a brainwave. He could realise Ellis's dream, using the mathematics of prime numbers to create a trapdoor function. Mathematically, multiplying two or more prime numbers together is easy. You can do it by hand with two 50-digit primes, getting a 99- or 100-digit result. The reverse, taking a number with a hundred digits and finding its prime factors, is far harder. The standard school method 'try possible factors in turn' is hopeless: there are too many possibilities. Cocks devised a trapdoor function based on the product of two large primes – what you get by multiplying them together. The resulting code is so secure that this product, though not the primes themselves, can even be made public. Decoding requires knowing the two primes *separately*, and that's the secret second key. Unless you know those two primes, you're stuck; knowing their product alone doesn't help. For instance, suppose I tell you I've found two primes whose product is

1,192,344,277,257,254,936,928,421,267,205,031,305,805,339,598,743,
208,059,530,638,398,522,646,841,344,407,246,985,523,336,728,666,069.

Can you find those primes?[41] A really fast supercomputer can do it, but a laptop would struggle. With more digits, even the supercomputer is stumped.

Anyway, Cocks's background was in number theory, and he devised a way to use such a pair of primes to create a trapdoor function – I'll explain how a bit later on, when we have the necessary concepts. It was so simple that at first *he didn't even write it down*. Later, he put the details in a report to his superiors. But no one could think of a way to use this method, not with the rudimentary computers of the time, so it was classified. It was also passed on to the US National Security Agency. Both organisations did see military potential, because even if the calculations were slow, you could use the public key system to send someone the key to some other entirely different code electronically. That's the main way this type of cipher is used today, in both military and civilian applications.

Britain's bureaucrats have a long and undistinguished track record of failing to spot huge moneyspinners – penicillin, the jet engine, DNA fingerprinting. In this case, though, they can take some consolation from patent law: in order to patent something, you have to give away what it is. At any rate, Cocks's revolutionary idea was filed away, rather like the scene at the end of *Raiders of the Lost Ark*, where the box containing the Ark of the Covenant is wheeled away into the depths of a gigantic, anonymous government warehouse, piled to the rafters with boxes that look just the same.

Meanwhile, come 1977, the identical method surfaced, reinvented independently and promptly published by three American mathematicians: Ronald Rivest, Adi Shamir, and Leonard Adleman. After them, we now call it the RSA cryptosystem. Finally, in 1997, the British security services declassified Cocks's work, which is how we now know he thought of it first.

*

Number theory enters cryptography as soon as we realise that any message can be represented by a number. For Caesar's cipher, that number is the position of a letter in the alphabet, which mathematicians prefer to run from 0 to 25 instead of 1 to 26, for reasons of algebraic convenience. So A is 0, B is 1, and so on up to Z = 25. Numbers outside this range can be converted into numbers that are inside it by adding or subtracting multiples of 26. This convention wraps the 26 letters round in a circle, so that after Z we go back to A. Caesar's cipher can then be condensed into a simple mathematical rule, indeed a formula:

$$n \rightarrow n + 3$$

The reverse process looks very similar:

$$n \leftarrow n + 3 \quad \text{or} \quad n \rightarrow n - 3$$

This is what makes the cipher symmetric.

We can invent new codes by changing the rules – by changing the formula. We just need a simple way to convert a message into a number, and two formulas: one to turn the plaintext message into ciphertext, and another to get it back again. Each formula has to reverse the other one.

There are lots of ways to convert plaintext into numbers. A simple one is to use 0–25 for each letter, and string these numbers together, padding out 0–9 as 00–09. So JULIUS would become 092011082018 (remember, A = 00). Maybe extra numbers are needed for space, punctuation, whatever. A rule that turns one number into another is called a number-theoretic function.

Wrapping numbers round in a circle is a standard number theorist's trick, called modular arithmetic. Pick a number – here 26. Now pretend that 26 is the same as 0, so the only numbers you need are 0–25. In 1801 Carl Friedrich Gauss, in his famous *Disquisitiones Arithmeticae* (Arithmetical Investigations) pointed out that in such a system you can add, subtract, and multiply numbers, obeying all the usual laws of algebra, without straying from the chosen range 0–25. Just do the usual calculation with ordinary numbers and then take the remainder on dividing the answer by 26. So, for example, $23 \times 17 = 391$, which is $15 \times 26 + 1$. The remainder is 1, so $23 \times 17 = 1$ in this unusual version of arithmetic.

The same idea works with 26 replaced by any other number; this number is called the *modulus*, and we can write (mod 26) to emphasise what's going on. So, more properly, we've calculated that $23 \times 17 = 1$ (mod 26).

What about division? If we divide by 17, and don't worry too much about what that means, we get

$$23 = 1/17 \quad (\text{mod } 26)$$

so dividing by 17 is the same as multiplying by 23. We can now invent a new code rule:

$$n \rightarrow 23n \quad (\text{mod } 26)$$

whose reverse is

$$n \leftarrow 17n \quad (\text{mod } 26)$$

This rule jumbles up the alphabet considerably, into the order

AXUROLIFCZWTQNKHEBYVSPMJGD

It's still a substitution code on the level of single letters, so it can easily be broken, but it makes the point that we can change the formula. It also illustrates the use of modular arithmetic, the key to large areas of number theory.

However, division can be trickier. Since $2 \times 13 = 26 = 0$ (mod 26), we can't divide by 13, otherwise we'd deduce that $2 = 0/13 = 0$ (mod 26), which is wrong. The same goes for division by 2. The general rule is that we can divide by any number that doesn't share any prime factor with the modulus. So 0 is ruled out, but that's no surprise: we can't divide by 0 in ordinary whole numbers. If the modulus is prime, we can divide by any number less than the modulus, except 0.

The advantage of modular arithmetic is that it gives the list of plaintext 'words' an algebraic structure. This opens up a wide variety of rules for transforming plaintext into ciphertext, and back again. What Cocks did, and later Rivest, Shamir, and Adleman, was to pick a very clever rule.

Encoding a message one letter at a time, using the same numbering for each letter, isn't terribly secure: whatever the rule, we have a substitution code. But if we divide the message into blocks, say ten letters long, or nowadays more like a hundred, and convert each block into a number, we have a substitution code on blocks. If the blocks are long enough, there's no distinctive pattern to the frequency with which any block occurs, so decoding by observing which numbers occur more often won't work any more.

*

Cocks and RSA derived their rules from the beautiful theorem that Fermat discovered in 1640, telling us how *powers* of numbers behave in modular arithmetic. In modern language, Fermat told his friend de Bessy that if n is prime then

$$a^n = a \ (\text{mod } n) \quad \text{or equivalently} \quad a^{n-1} = 1 \ (\text{mod } n)$$

for any number a. 'I would send you a demonstration of it, if I did not fear going on for too long,' Fermat wrote. Euler supplied the missing proof in 1736, and in 1763 he published a more general theorem that applies when the modulus isn't prime. Now a and n mustn't share a common factor, and the power $n - 1$ in the second version of the formula is replaced by Euler's 'totient' function $\varphi(n)$. We don't need to know what that is,[42] but we do need to know that if $n = pq$ is the product of two primes p and q then $\varphi(n) = (p - 1)(q - 1)$.

The RSA cryptosystem proceeds like this:

- Find two large primes p and q.
- Compute the product $n = pq$.
- Compute $\varphi(n) = (p - 1)(q - 1)$. Keep this secret.
- Choose e having no prime factor in common with $\varphi(n)$.
- Calculate d so that $de = 1 \pmod{\varphi(n)}$.
- The number e can be made public. (This gives very little useful information about $\varphi(n)$, by the way.)
- Keep d private. (This is vital.)
- Let r be a plaintext message, coded as a number modulo n.
- Convert r to the ciphertext $r^e \pmod{n}$. (This rule can also be made public.)
- To decode r^e, raise it to the power $d \pmod{n}$. (Remember, d is secret.) This gives $(r^e)^d$, which equals r^{ed}, which equals r by Euler's theorem.

Here the encoding rule is 'take the eth power':

$$r \rightarrow r^e$$

and the decoding rule is 'take the dth power':

$$s \rightarrow s^d$$

Some mathematical tricks, which I won't go into, make it possible to carry out all of these steps quickly (on today's computers), provided you know p and q *separately*. The sting in the tail is that if you don't, then knowing n and e doesn't greatly help in calculating d, which you need to decode the message. Essentially, you need to find the prime factors p and q of n, which we've seen is (apparently) much harder than multiplying p and q to obtain n.

In other words, 'raise to the power e' is the required trapdoor function.

Currently, everything here can be done in a minute or so on a laptop for, say, 100-digit primes p and q. One pleasant feature of the RSA system is that as computers become more powerful, all you need to do is make p and q bigger. The same method works.

One disadvantage is that RSA, though entirely practical, is too slow to be used routinely for the entire content of every message. The main practical application is to use RSA as a secure way to transmit a secret key for some entirely different cipher system – one that's far faster to implement and is secure as long as no one knows the key. So RSA solves the problem of key distribution, which has bedevilled cryptography since its earliest days. One reason Enigma was broken was because certain settings on the Enigma machine were distributed to operators at the start of each day in an insecure manner. Another common application is to verify an electronic signature, that is, a code message establishing the sender's identity.

Cocks's boss Ralph Benjamin, Chief Scientist, Chief Engineer, and Superintending Director at GCHQ, was very much on the ball, and spotted this possibility. He wrote in a report: 'I judged it most important for military use. In a fluid military situation you may meet unforeseen threats or opportunities. If you can share your key rapidly and electronically, you have a major advantage over your opponent.' But computers weren't up to the task at the time, and the British government missed what, with hindsight, was a huge opportunity.

*

Mathematical techniques seldom solve practical problems 'off the peg'. Like everything else, they generally need to be adapted and tweaked to overcome various difficulties. This goes for RSA: it's not quite as simple as I've just described. In fact, a number of fascinating theoretical questions for mathematicians emerge as soon as we stop admiring the idea and think about what could go wrong.

It's not hard to show that calculating $\varphi(n)$, without knowing its prime factors p and q, is just as hard as finding p and q themselves. In fact, that seems to be the only way to do it. So the big question is: how hard is prime factorisation? Most mathematicians think that it's extremely difficult, in a technical sense: any factorisation algorithm has a running time that grows explosively with the number of digits in the product pq. (Incidentally, the reason for using just two primes, rather than, say, three, is that this is the most difficult case. The more prime factors a number has, the easier it is to find one of them. Divide it out, and the number is now a lot smaller, so it's easier to find the rest as well.) However, no one can currently *prove* that prime factorisation is hard. No one has a clue how to get started on such a proof. So the security of the RSA method rests on an unproved conjecture.

The other questions and pitfalls involve fine details of the method. Poor choices of the numbers used can render RSA vulnerable to ingenious attacks. For example, if e is too small, then we can determine the message r by taking the eth root of the ciphertext r^e, considered as an ordinary number – that is, not mod n. Another potential flaw arises if the same message is sent to e recipients using the same power e, even if p and q are different for each of them. A lovely result called the Chinese Remainder Theorem can then be brought to bear, disclosing the plaintext message.

RSA as described is also semantically insecure, which means that in principle it might be cracked by encoding lots of different plaintext messages and trying to match the result to the ciphertext you want to break. Basically, by trial and error. This may not be practical for long messages, but if lots of short messages are being sent it might become so. To avoid this, RSA is modified by padding the message with additional digits, according to some specific but random scheme. This makes the plaintext longer and avoids sending the same message many times.

Another method for breaking an RSA code exploits not a mathematical flaw, but a physical feature of the computer. In 1995 the cryptography entrepreneur Paul Kocher observed that if the codebreaker knows enough about the hardware being used, and can measure how long it takes to decode several messages, then the secret key d can easily be deduced. Dan Boneh and David Brumley demonstrated a practical version of this attack in 2003 for messages sent over a conventional network using the standard SSL (Secure Sockets Layer) protocol.

The existence of mathematical methods that can *sometimes* factorise a large number very fast implies that the primes p and q must be chosen to satisfy some restrictive conditions. They should not be too close together, or else a factorisation method that goes right back to Fermat applies. In 2012 a group under Arjen Lenstra tried this out on millions of public keys extracted from the Internet, and were able to crack one in five hundred of them.

The big game-changer would be a practical quantum computer. These machines, still in their infancy, use quantum bits in place of the usual binary digits 0 and 1, and in principle can perform gigantic computations, such as factorising huge numbers, with unprecedented speed. I'll postpone further discussion to later in this chapter.

*

The RSA system is only one of a number of ciphers based on number theory, or its close relative combinatorics, a method for counting how many ways some arrangement can be achieved without listing all the possibilities. To convince you that the mathematical well has not yet run dry as far as cryptography is concerned, I'll describe an alternative cipher system, which exploits one of the deepest and most exciting areas of today's number theory. This area is about 'elliptic curves', which among other things are central to Andrew Wiles's epic proof of Fermat's Last Theorem.

Number theory has moved on since the time of Fermat and Euler. So has algebra, the emphasis shifting from symbolic representation of unknown numbers to general properties of symbolic systems defined by specific rules. These two areas of research overlap substantially. Some fascinating ideas about secret ciphers have emerged from a combination of two specialist areas of algebra and number theory: finite fields and elliptic curves. To understand what's involved, we first need to know what these things are.

We saw that in arithmetic to some modulus, it's possible to add, subtract, and multiply 'numbers' while obeying the usual algebraic rules. To avoid getting sidetracked, I didn't actually say what those rules are, but typical examples are the commutative law $ab = ba$ and the associative law $(ab)c = a(bc)$. These apply to multiplication, and there are similar laws for addition. The distributive law $a(b+c) = ab + ac$ also holds, and there are simple rules involving 0 and 1, such as $0 + a = a$ and $1a = a$. Any system that obeys these laws is called a *ring*. If division is also possible (except by 0) and the standard rules apply, we get a *field*. The names are traditional, imported from German, and basically just mean 'some sort of collection of stuff that obeys the specified rules'. The integers modulo 26 form a ring, known as $\mathbb{Z}_{26}$. We saw that there are problems when dividing by 2 or 13, so it's not a field. I said (without indicating why) that integers modulo a prime number don't have such problems, so $\mathbb{Z}_2$, $\mathbb{Z}_3$, $\mathbb{Z}_5$ $\mathbb{Z}_7$, and so on – integers modulo 2, 3, 5, 7 – are all fields.

The ordinary whole numbers go on forever: they form an infinite set. In contrast, systems like $\mathbb{Z}_{26}$ and $\mathbb{Z}_7$ are finite. The first one comprises only the numbers 0–25, the second 0–6. The first is a finite ring, the second a finite field. It's really quite remarkable that finite systems can obey so many of the rules of algebra without any logical inconsistencies. Finite number systems, if not too big, are very well suited to computer calculations, because these can done *exactly*. It's therefore no surprise that a variety of codes are based on finite fields. Not just cryptographic ciphers, to ensure secrecy, but error-detecting and error-correcting codes, to ensure that messages are received without errors arising from random 'noise', such as electrical interference. A whole new area of mathematics, coding theory, addresses such issues.

The simplest finite fields are $\mathbb{Z}_p$, the integers modulo a prime p. That they form a field was known (though not by that name) to Fermat. The French revolutionary Évariste Galois, killed in a tragic duel aged 20, proved that these aren't the only finite fields. He found them all: there's one finite field for each prime power p^n, and it contains exactly p^n different 'numbers'. (Warning: if n is bigger than 1, this field is not the integers modulo p^n.) So there are finite fields with 2, 3, 4, 5, 7, 8, 9, 11, 13, 16, 17, 19, 23, 25, … elements, but not with 1, 6, 10, 12, 14, 15, 18, 20, 21, 22, 24, … elements – a very curious theorem.

Elliptic curves (which are related only very indirectly to ellipses) originated in a different area, the classical theory of numbers. Around the year 250 the ancient Greek mathematician Diophantus of Alexandria wrote a text about solving algebraic equations using whole (or rational) numbers. For instance, the famous 3–4–5 triangle has a right angle, thanks to Pythagoras, because $3^2 + 4^2 = 5^2$. These numbers therefore solve the Pythagorean equation $x^2 + y^2 = z^2$. One of Diophantus's theorems shows how to find all solutions of this equation in fractions, and in particular in whole numbers. This general area, solving equations in rational numbers, became known as *Diophantine equations*. The restriction to rational numbers changes the game; for instance $x^2 = 2$ can be solved in real numbers but not in rational ones.

One of Diophantus's problems is: 'Divide a given number into two numbers whose product is a cube minus its side.' If the original number is a, we split it into Y and $a - Y$, and want to solve

$$Y(a - Y) = X^3 - X$$

Diophantus examined the case when $a = 6$. A suitable change of variables (subtract 9, change Y to $y + 3$ and X to $-x$) converts this equation into

$$y^2 = x^3 - x + 9$$

He then derived the solution $X = 17/9$, $Y = 26/27$.

Remarkably, similar equations turned up in geometry, when mathematicians tried to use analysis (advanced calculus) to calculate the arc length of a segment of an ellipse. Indeed, this is how the name 'elliptic curve' arose. They knew how to answer the analogous question for a circle, using calculus. The problem then reduces to finding the integral of a function involving the square root of a quadratic polynomial, and this can be done using (inverse) trigonometric functions. The same method applied to an ellipse leads to the integral of a function involving the square root of a cubic polynomial, and after some fruitless experimentation, it became clear that a new class of functions was needed. These functions turned out to be rather pretty, though complicated, and they were named elliptic functions because of their connection with the arc length of an ellipse. The square root of a cubic polynomial is a solution y of the equation

$$y^2 = x^3 + ax + b$$

(any x^2 term on the right can be transformed to zero). In coordinate geometry this equation defines a curve in the plane, so such curves (and their algebraic version as an equation) became known as 'elliptic curves'.

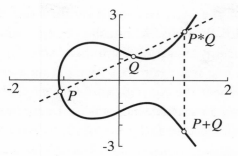

To 'add' two points P and Q on an elliptic curve, join them with a line meeting the curve at a third point $P*Q$. Then reflect in the x-axis to get $P+Q$.

When the coefficients are integers, we can consider the equation in modular arithmetic, say in $\mathbb{Z}_7$. Every solution in ordinary integers leads to one in arithmetic modulo 7. Since this system is finite, we can use trial and error. For Diophantus's equation $y^2 = x^3 - x + 9$ we quickly discover that the only solutions (mod 7) are:

$$x = 2, y = 2 \quad x = 2, y = 5$$
$$x = 3, y = 1 \quad x = 3, y = 6$$
$$x = 4, y = 3 \quad x = 4, y = 4$$

These solutions have implications for any solution in ordinary integers: it must reduce modulo 7 to one of these six. The same goes for rational solutions, provided the denominator is not a multiple of 7 – those are forbidden because in $\mathbb{Z}_7$ such a denominator becomes 0. Change 7 to some other number and we get more information about the form of any hypothetical rational solution.

Now we're looking at elliptic curves – equations – over finite rings and fields. The geometric image of a curve is not really applicable, since we just have a finite set of points, but it's convenient to use the same name. The picture shows a typical shape, and an extra feature, known to Fermat and Euler, that intrigued mathematicians of the early twentieth century. Given two solutions, we can 'add' them to get another one, as shown in the picture. If the solutions are rational numbers, so is their sum. Not just 'buy two, get one free', but 'buy two, get an awful lot free', because you can repeat the construction. Occasionally this leads you back to where you started, but mostly it generates infinitely many different solutions. In fact, the solutions have a nice algebraic structure: they form a group, the Mordell–Weil group of the elliptic curve. Louis Mordell proved its basic properties and André Weil generalised them, and 'group' means that addition obeys a short list of simple rules. This group is commutative, meaning that $P + Q = Q + P$, which is obvious from the picture since the line through P and Q is also the line through Q and P. The existence of such a group structure is unusual; most Diophantine equations are less obliging. Many have no solutions at all, some have just a few, and it's difficult to predict which kind you're looking at. In fact, elliptic curves are the focus of intensive research, for this and other reasons. Andrew Wiles proved a deep conjecture about elliptic curves as a key step in his proof of Fermat's Last Theorem.

*

The group structure of an elliptic curve also interests cryptographers. It's usually considered as a form of 'addition' of solutions, although the formula is far more complicated, because it's commutative, and the symbol + has become traditional in the theory of commutative groups. In particular, if we have a solution (x, y), which we can think of as a point P in a plane, then we can generate solutions $P + P$, $P + P + P$, and so on. It's natural to call these $2P$, $3P$, and so on.

In 1985 Neal Koblitz and Victor Miller independently realised that you can use the group law on an elliptic curve to create a cipher. The idea is to work in some finite field, with a large number of elements. To encode P, we work out kP for some very large integer k, which is easy with a computer, and call the result Q. To reverse this process, we must start with Q and find P – in effect, dividing by k. Because of the complexity of the group formula, this reversed calculation is very hard, so we have invented a new kind of trapdoor function, hence a public key cryptosystem. It's known as Elliptic Curve Cryptography, or ECC. Just as RSA can be applied using many different primes, ECC can be applied using many different elliptic curves, over many different finite fields, with different choices of P and of the multiplier k. Again, there's a secret key that allows rapid decoding.

The advantage turns out to be that a smaller group leads to a cipher that's just as secure as an RSA code based on much larger prime numbers. So the elliptic curve cipher is more efficient. Putting a message into code, and decoding it provided you know a secret key, is quicker and simpler. Breaking the code if you don't know the key is just as hard. In 2005 the US National Security Agency recommended that research in public key cryptography should move on to the new area of elliptic curves.

As for the RSA system, there's no rigorous proof that ECC is secure. The range of possible attacks is similar to that for RSA.

There's a lot of interest in the moment in cryptocurrencies, which are monetary systems not controlled by conventional banks, although the banks are getting interested too. Always alert for any new way to make money, the banks. The best-known cryptocurrency is Bitcoin. The security of bitcoins is ensured by a technique called the blockchain, which is an encoded record of all transactions involving that particular 'coin'. New bitcoins are brought into existence by 'mining', which essentially means performing a huge number of otherwise pointless computations. Bitcoin mining uses up substantial quantities of electricity to no useful purpose except enriching a few individuals. In Iceland, where electricity is very cheap thanks to thermal generation from underground steam, bitcoin mining uses more electricity than everyone's homes combined. It's hard to see how this activity helps to combat global heating and the climate crisis, but there you go.

Bitcoin and many other cryptocurrencies use one particular elliptic curve, known by the catchy name secp256k1. Its equation, $y^2 = x^3 + 7$, is far more catchy, and this seems to be the main reason it was chosen. Encoding via secp256k1 is based on a point on the curve given by

$$x = 55066263022277343669578718895168534326250\\6034537775941755001873603891167292240$$
$$y = 32670510020758816978083085130507043184471273380659243924327593890433575733748242424$$

This illustrates the gigantic integers involved in practical implementations of ECC.

*

I've said several times that the security of the RSA system rests on the unproved presumption that prime factorisation is difficult. Even if that's correct, and it very likely is, there might be other ways to compromise the cipher's security, and the same goes for all of the classical public key encryption schemes. One potential way this could happen is if someone invents a computer that's much faster than anything currently available. Today, this new type of security threat is looming on the horizon: the quantum computer.

A classical physical system exists in a specific state. A coin on a table is either heads or tails. A switch is either on or off. A binary digit (or 'bit') in a computer memory is either 0 or 1. Quantum systems aren't like that. A quantum object is a wave, and waves can be piled on top of each other, which technically is called a superposition. The state of a superposition is a mixture of the states of the components. The celebrated (indeed, notorious) case of Schrödinger's cat is a vivid example: by some chicanery with a radioactive atom and a flask of poison gas, along with a cat in an impermeable box, the quantum state of the unfortunate animal can be a superposition of 'alive' and 'dead'. A classical cat must be one or the other, but a quantum puss can be both at the same time.

Until you open the box.

Then the cat's wave function 'collapses' to just one of the classical states. Either it's alive, or it's dead. Curiosity (opening the box) killed the cat. Or not.

I don't want to go into the controversial and often heated debate about whether quantum states really would work like this for cats.[43] What matters here is that the mathematical physics works beautifully for simpler objects, already being used to make rudimentary quantum computers. In place of the bit, either 0 or 1, we have the qubit, both 0 and 1 at the same time. A classical computer, the kind you and I have on our desk, in our bag, or in our pocket, handles information as a sequence of 0's and 1's. It actually uses quantum effects to do that, so small is the scale of today's computer circuits, but the upshot is that the computations correspond to classical physics. Engineers building classical computers work very hard to ensure that 0 remains 0 and 1 remains 1, and never the twain shall meet. The classical cat must either be alive or dead. So a register of (say) 8 bits can store a single sequence like 01101101 or 10000110.

In a quantum computer, exactly the opposite prevails. A register of 8 qubits can store both of those at the same time, along with the other 254 possible 8-bit sequences. Moreover, it can do sums with all 256 possibilities *simultaneously*. It's like having 256 computers instead of just one. The longer the sequences, the more the number of possibilities explodes. A 100-bit register can store a single sequence of 100 bits. A 100-qubit register can store, and manipulate, all 10^{30} possible sequences of 100 bits. This is 'parallel processing' on a massive scale, and it's what's got so many people excited about quantum computers. Instead of doing 10^{30} calculations one at a time, you do them *all at once*.

In principle.

In the 1980s Paul Benioff proposed a quantum model of the Turing machine, the theoretical formulation of classical computing. Soon after, the physicist Richard Feynman and the mathematician Yuri Manin pointed out that a quantum computer might be able to perform huge numbers of calculations in parallel. A major breakthrough on the theoretical aspect came in 1994 when Peter Shor invented a very fast quantum algorithm to factor large numbers into primes. This shows that the RSA cryptosystem is potentially vulnerable to attack by an enemy using a quantum computer, but more importantly, it shows that a quantum algorithm can massively outperform a classical one on a sensible, uncontrived, problem.

In practice, the obstacles to building a practical quantum computer are huge. Tiny disturbances from external sources, or even just the vibration of molecules that we call heat, cause a superposed state to 'decohere', that is, break up, very fast. To mitigate this problem, the machine currently has to be cooled to very near absolute zero, $-273°C$, which requires supplies of helium-3, a rare by-product of nuclear reactions. Even this can't prevent decoherence occurring; it just slows it down. So every calculation has to be embellished with an error-correcting system that spots disruption from outside sources and puts qubit states back where they ought to be. The Quantum Threshold Theorem tells us that this technique works provided it can correct errors faster than decoherence introduces them. As a ballpark estimate, the error rate for each logic gate must be at most one per thousand.

Error correction also brings a penalty: it requires more qubits. For example, to factor a number that can be stored in n qubits using Shor's algorithm, the calculation runs in a time roughly proportional to something between n and n^2. With error correction, which in practice is essential, this becomes more like n^3. For a 1,000-qubit number, error correction multiplies the running time by a thousand.

Until very recently, no one had built a quantum computer with more than a few qubits. In 1998 Jonathan A. Jones and Michele Mosca used a 2-qubit device to solve Deutsch's problem. This stems from work of David Deutsch and Richard Jozsa in 1992. It's a quantum algorithm that runs exponentially faster than any conventional algorithm, it always gives an answer, and that answer is always correct. The problem it solves is this. We're given a hypothetical device, an *oracle*, which implements some Boolean function: one that turns any n-digit bit string into either 0 or 1. Mathematically, the oracle *is* that function. We're also told that the Boolean function takes a value of either 0 everywhere, or 1 everywhere, or 0 on precisely half of the bit strings and 1 on the other half. The problem is to determine which of these three cases occurs, by applying the function to bit strings and observing the result. Deutsch's problem is deliberately artificial, proof of concept rather than practical. Its merit is that it exhibits a specific problem for which a quantum algorithm provably outstrips any conventional one. Technically, it proves that the complexity class EQP (exact polynomial-time solutions on a quantum computer) is different from class P (exact polynomial-time solutions on a conventional computer).

The year 1998 also saw a 3-qubit machine, and in 2000 machines with 5 and 7 qubits appeared. In 2001 Lieven Vandersypen and coworkers[44] implemented Shor's algorithm using seven spin-1/2 nuclei in a specially synthesised molecule as quantum bits, which can be manipulated with room temperature liquid-state nuclear magnetic resonance techniques, to find the prime factors of the integer 15. Most of us can do that in our heads, but this was an important proof of concept. By 2006, researchers had reached 12 qubits, with claims of 28 in 2007 by a company called D-Wave.

*

While this was going on, researchers greatly increased the length of time that a quantum state could be made to persist, before decohering. In 2008 a qubit was stored for more than a second in an atomic nucleus. By 2015 the lifetime was six hours. Comparison of these times is difficult because different devices used different quantum methods, but progress was impressive. In 2011 D-Wave announced that it had constructed a commercially available quantum computer, D-Wave One, with a 128-qubit processor. By 2015 D-Wave claimed to have passed 1,000 qubits.

The early response to D-Wave's claims was sceptical. The device's architecture was unusual, and some questioned whether it was a true quantum computer rather than a fancy classical computer using quantum-related gadgetry. In tests it outperformed off-the-shelf computers on useful tasks, but it had been designed specifically for those tasks, while the classical computers it was competing against had not. Its advantages seemed to disappear when classical devices were designed specifically for the tasks concerned. The controversy continues, but D-Wave's machines are in use and working well.

A key goal of the research is quantum supremacy: making a quantum device that outperforms the best classical computers on at least one calculation. In 2019, a team at Google AI published a paper in *Nature* with the title 'Quantum supremacy using a programmable superconducting processor'.[45] They announced that they had built a quantum processor called Sycamore with 54 qubits, but one had failed, reducing the number to 53. They had used it to solve, in 200 seconds, a problem that would take a classical computer 10,000 years.

This claim was immediately challenged on two grounds. One was that the computation could probably be done in a shorter time by a classical machine. The other was that the problem Sycamore solved is rather contrived: sampling the output of a pseudorandom quantum circuit. The circuit layout connects components randomly, the aim being to calculate the probability

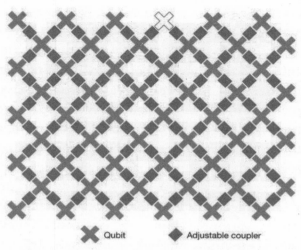

Architecture of the Sycamore quantum processor.

distribution of a sample of the possible outputs. Some outputs turn out to be far more likely than others, so this distribution is very complicated and non-uniform. A classical calculation grows exponentially with the number of qubits. Nevertheless, the team had succeeded in its primary aim: to show that there's no practical obstacle to creating a quantum computer that can beat classical ones at *something*.

A question that immediately springs to mind is: how do you tell whether the answer is correct? You can't wait 10,000 years for a classical computer to solve the problem, and you can't simply believe the result without checking it. The team tackled this using a method called cross-entropy benchmarking, which compares probabilities of specific bit strings to theoretical ones calculated on a classical computer. This provides a measure of how likely the result is to be right; the conclusion was that it's accurate to within 0·2% with very high ('five-sigma') probability.

Despite all this progress, most experts think that a practical quantum computer is still a long way off. Some remain unconvinced that it can ever be achieved. The physicist Mikhail Dyakonov wrote:

> The number of continuous parameters describing the state of such a useful quantum computer at any given moment must be ... about 10^{300}. ... Could we ever learn to control the more than 10^{300} continuously variable parameters defining the quantum state of such a system? My answer is simple. *No, never.*

*

Dyakonov could be right, but others disagree. Either way, the mere possibility that someone – probably a massive research team funded by a government or major corporation – *might* construct a quantum computer is enough to give many countries' security services and financial industries nightmares. Enemy forces would be able to decrypt military messages; criminals could destroy Internet commerce and banking. So theoreticians have turned their attention to what cryptography would look like in a post-quantum world, trying to get ahead of the game and restore secure communications.

The good news is that what a quantum computer can break it can also render unbreakable. This requires new cryptographic methods, using quantum computation to create new codes that even a quantum computer can't break. This would entail a new way of thinking about the underlying mathematics. An interesting feature is that much of it still uses number theory, though of a more modern vintage than Fermat.

The potentially imminent advent of quantum computers has set off a wave of research devising encryption methods that can't be broken by a quantum computer. Recently the National Institute of Standards and Technology (NIST) has started a programme of Post Quantum Cryptography, aimed at identifying classical cryptosystems that are at risk and finding new ways to combat their vulnerabilities. In 2003 John Proos and Christof Zalka[46] estimated the vulnerability of the RSA system and elliptic curve cryptography to a quantum computer running Shor's algorithm. Martin Roetteler and coworkers[47] updated their results in 2017. They proved that for an elliptic curve over a finite field with q elements, where q is roughly 2^n, RSA is vulnerable to a quantum computer with $9n + 2 \log_2 n + 10$ qubits and a circuit of at most $448n^3 \log_2 n + 4090n^3$ Toffoli gates. A Toffoli gate is a special kind of logic circuit, from which circuits can be constructed to carry out any logical function. Moreover, it's reversible: you can work backwards from the output to deduce the inputs. The current standard for RSA is to use a number with 2048 bits, that is, about 616 decimal places. The team estimated that 2048-bit RSA will be vulnerable to a quantum computer with $n = 256$, and elliptic curve encryption would be vulnerable to a quantum computer with $n = 384$.

Identifying vulnerabilities is all very well, but the big question is what to do to protect against them. This requires entirely new cryptographic methods. The general idea is the same as always: base the encryption method on hard mathematical problems with some sort of easy back door. But now 'hard' means 'hard for a quantum computer'. Right now, four main classes of problems of this kind have been identified:

- Random linear error-correcting codes
- Solving systems of nonlinear equations over large finite fields
- Finding short vectors in high-dimensional lattices
- Finding paths between random vertices of random-looking graphs.

Let's take a quick look at the fourth of these, which involves the newest ideas and some very advanced mathematics.

For practical purposes, we work with a graph that has about 10^{75} vertices and similarly large number of edges. The code depends on finding a path through this graph between two specific vertices. This is a type of TSP, and is comparably hard. To create a back door, the graph must have some hidden structure that makes the solution easy. The central idea is to use cutely named supersingular isogeny graphs, or SIGs. These are defined using elliptic curves with special properties, said to be supersingular. The vertices of the graph correspond to *all* supersingular elliptic curves over the algebraic closure of a finite field with p elements. There are about $p/12$ such curves.

An isogeny between two elliptic curves is a polynomial map from one to the other that preserves the Mordell–Weil group's structure. We use isogenies to define the edges of the graph. To do so, take a second prime q. The edges of this graph correspond to degree-q isogenies between the two elliptic curves corresponding to the ends of that edge. Exactly $q + 1$ edges emanate from each vertex. These graphs are 'expander' graphs, which means that random walks starting at any vertex diverge rapidly as the walk proceeds, at least for a large number of steps.

An expander graph can be used to create a hash function, which is a Boolean function from n-bit strings to m-bit strings, where m is much less than n. Alice can use a hash function to convince Bob that she knows some particular n-bit string, also known to Bob, *without divulging what it is*. Namely, she forms the much shorter hash function of that string and sends Bob that. He calculates the hash function of his string, and compares.

Two conditions are needed for this method to be secure. One is a trapdoor condition called preimage resistance: it's not computationally feasible to invert the hash function and work out an n-bit string that gives that hash. There will generally be lots, but the point is that in practice you can't find *any*. The other desideratum is a collision-resistant hash function, which means that it's not computationally tractable to find two distinct n-bit strings with the same m-bit hash. What this means is that if an eaves-dropper Eve overhears the conversation, the hash that Alice sends doesn't help her work out what the original n-bit string was.

Given two primes p and q with some extra technical conditions, we can exploit this idea by constructing the corresponding SIG and use its expander properties to define a preimage-resistant, collision-resistant hash function. This can then be used to create a highly secure code. Breaking the code requires computing lots of isogenies between elliptic curves. The best quantum algorithm for one such calculation runs in time $p^{1/4}$. Make p and q large enough (the mathematics tells us how large) and you have a cryptosystem that even a quantum computer can't break.

All very technical. I don't expect you to understand the details; for a start, I haven't told you most of them. But I hope you get the message that very advanced and abstract mathematics, to do with algebraic geometry over finite fields, might be just what we need to protect our personal, commercial, and military communications against eavesdroppers armed with currently hypothetical but possibly soon all-too-real quantum computers.

Hardy's beloved number theory has become far more useful than he imagined. But some of today's applications would have disappointed him. Maybe we should apologise to him.

6

The Number Plane

The Divine Spirit found a sublime outlet in that wonder of analysis, that portent of the ideal world, that amphibian between being and not-being, which we call the imaginary root of negative unity.

Gottfried Wilhelm Leibniz, *Acta Eruditorum*, 1702

We are currently in the midst of a second quantum revolution. The first quantum revolution gave us new rules that govern physical reality. The second quantum revolution will take these rules and use them to develop new technologies.

Jonathan Dowling and Gerard Milburn, *Philosophical Transactions of the Royal Society*, 2003

There's been a lot of activity in our part of the city of Coventry in recent months. White vans have been parked all over the place at the roadside, often accompanied by trucks laden with shovels and wheelbarrows. Mini-diggers have patrolled the streets on caterpillar tracks, scraping trenches along pavements, across roads, through gardens, and the newly laid tarmac rambles over the landscape like slime trails left by dog-sized snails. Men in hi-vis jackets have appeared, apparently being swallowed up by holes in the ground where manhole covers have opened up. Coils of cable decorate verges and are propped up against hedges, waiting to be sucked into the manholes. Puzzled engineers sit under awnings in the rain, fiddling with thousands of colour-coded wires inside large metal boxes.

The vans have a message on the side that explains all this activity. *Superfast Fibre Broadband is in your area.*

The UK's city centres had this marvel of modern communications installed years ago, but our house is out in the boondocks. A company once declined to visit because it was so far out – all of four miles. To be fair, the city boundary is only a few hundred yards away. It's more expensive to put in the cables, and the population density is lower because just over the boundary the land is mainly farmers' fields. No easy expansion in that direction at marginal cost. We just haven't been an attractive proposition. But finally, after the government started putting pressure on the telecoms corporations, there's been an all-out push to get optical fibre connections to all urban areas and most rural ones. Instead of hopelessly watching areas with the densest populations repeatedly being upgraded to ever-faster services, because those areas are more profitable, the rest of the country is finally catching up. Or, at least, not falling even further behind.

In an age where virtually every activity has been hived off to the Internet, fast broadband has changed from a luxury to an essential. Maybe not as vital as water or electricity, but at least as necessary as the telephone. The advanced electronics that's driving the computer revolution and rapid worldwide communications has turned the 2020s into a world that would have seemed utterly alien in the 1990s. And it's only just getting started. Increased supply is creating an explosive growth in demand. The days when phone lines were made of copper and carried conversations are fast disappearing – and even those only worked, in recent years, because of cunning electronic and mathematical trickery to increase capacity. Today, communications cables carry far more data than they do conversations. That's why optical fibre has come to the fore.

Within a few decades, fibre will be as outmoded as the horse and cart. Future advances, allowing far larger amounts of data to be transmitted with breathtaking speed, are in the pipeline. Some already exist. The classical physics of electricity and magnetism remains fundamental, but electronic engineers are increasingly turning to the bizarre world of the quantum to build the next generation of communication devices. Underpinning both the classical physics and the quantum mechanics upon which all these developments rely is one of the most curious mathematical inventions ever made. It can be traced all the way back to ancient Greece, gained a tenuous foothold during the Italian Renaissance, and came to full flower in the nineteenth century, when it rapidly took over most of mathematics. It was widely used long before anyone really understood what it was.

I call it an invention rather than a discovery because it didn't draw inspiration from the natural world. If it was 'out there' waiting to be found, then 'there' was a very strange place, the world of human imagination and the obligations of logic and structure. It was a new kind of number, so new that it was given the name 'imaginary'. That name remains in use today, and imaginary numbers remain utterly alien to most of us, even though our lives increasingly depend on them.

You've heard of the number line.

Meet the number plane.

*

To understand how this strange development came about, and why, we must first take a look at the traditional number types. Numbers are so ordinary, so familiar, that it's easy to underestimate their subtleties. We know that two plus two makes four, and that five times six makes thirty. But what *are* 'two', 'four', 'five', 'six', and 'thirty'? They're not the words: different languages use different words for the same numbers. They're not the symbols 2, 4, 5, 6, 30: different cultures use different symbols. In the binary notation used in computing, these numbers are represented by 10, 100, 101, 110, and 11110. Anyway, what is a symbol?

It was all much simpler when a number was viewed as a straightforward description of nature. If you owned ten sheep, the number ten was a statement of how many sheep you owned. If you sold four of them, you would have six left. Numbers were basically an accounting device. But as mathematicians started using numbers in ever more esoteric ways, this pragmatic view began to look rather shaky. If you don't know what numbers are, how can you be sure your calculations never contradict each other? If a farmer counts the same flock of sheep twice, will she necessarily get the same answer? For that matter, what do we mean by 'count'?

In the 1800s nitpicking questions of this kind came to a head, because mathematicians had enlarged the number concept several times. Each new version incorporated what had gone before, but the link to reality was becoming increasingly indirect. First on the scene were the 'natural', 'whole', or 'counting' numbers 1, 2, 3, ... Next, fractions like 1/2, 2/3, or 3/4. At some point zero sneaked in. Up to that point, the correspondence with reality was fairly direct: take two oranges and another three oranges, and count them to verify that the total is five oranges. With the aid of a kitchen knife, I can show you half an orange. Zero oranges? An empty hand.

Even here, there are difficulties. Half an orange isn't exactly a *number* of oranges. It's not an orange at all, just a bit of one. There are many ways to cut an orange in half, and they don't all look the same. It's simpler with lengths of string, provided we cut them in the obvious way and don't do something silly like splitting strands lengthwise. Now it's all simple again. A piece of string has half the length of another one if two copies of the first, joined end to end, have the same length as the second. Fractions work best for *measuring* things. The ancient Greeks found measurements easier to handle than number symbols, so Euclid reversed this idea. Instead of using a number to measure the length of a line, he used the line to represent the number.

The next step, negative numbers, is trickier, because we can't show anyone minus four oranges. It's easier using money, where a negative number can be interpreted as a debt. All this was understood in China around AD 200, the first known source being the *Jiuzhang suanshu* (Nine Chapters on the Mathematical Art), but the idea is undoubtedly older. When numbers are associated with measurements, other interpretations of negative values arise naturally. For instance, a negative temperature can be interpreted as a temperature below zero, whereas a positive one is above zero. In some circumstances, a positive measurement lies to the right of some point, while a negative one lies to the left, and so on. Negative is opposite to positive.

Nowadays, mathematicians make a big fuss about the distinctions between these types of number system, but for ordinary users they're all variations on the same theme: *numbers*. We're happy to go along with this rather naive convention because the same rules of arithmetic work in all of these systems, and because each new type of number just extends the old system without changing what we already knew. The advantage of enlarging the number concept is that each extension makes it possible to do 'sums' that were previously impossible. In whole numbers we can't divide 2 by 3; in fractions we can. In whole numbers we can't subtract 5 from 3; in negative numbers we can. All of which makes mathematics *simpler*, because you can stop worrying about whether certain arithmetical operations are permitted.

*

Fractions can divide things up as finely as we wish. We can divide a metre into millimetres, one thousandth the size, or into micrometres, one millionth the size, or nanometres one billionth the size, and so on. We run out of names long before we run out of zeros. For practical purposes, there are always small errors of measurement, so fractions are all we need. In fact, we can get away with using only fractions whose denominator is a power of ten – look at any electronic calculator. But for vital theoretical purposes, and to keep mathematics neat and tidy, fractions proved inadequate.

The ancient Greek cult of the Pythagoreans believed that the universe runs on numbers, a view that's still prevalent in frontier physics, albeit in a more sophisticated way. The only numbers they recognised were whole numbers and positive fractions. So their belief system was shocked to the core when one of them discovered that the length of the diagonal of a square is not an exact fraction of the length of its side. This discovery led to so-called 'irrational' numbers, in this case the square root of 2. In a complicated historical development running from China in the fourth century BC to Simon Stevin in 1585, such numbers came to be represented as decimals:

$$\sqrt{2} = 1 \cdot 414,213,562,373,095,048, \ldots$$

Because this number is irrational, it must go on *forever*, without stopping with just zeros. It can't even repeat the same block of digits over and over again, like 1/3, which in decimals is $0 \cdot 333333333\ldots$. It's an 'infinite decimal'. We can never write it down in full, but conceptually we can pretend this is possible, because in principle we can write down as many of its digits as we wish.

Despite the need to appeal to an infinite process, infinite decimals have very pleasant mathematical properties, and in particular they provide *exact* representations for geometric lengths such as $\sqrt{2}$, which would otherwise not have numerical values at all. Infinite decimals came to be called 'real' numbers, because they were (idealised) measurements of real quantities such as length, area, volume, or weight. Each successive digit represents a multiple of a basic size that divides by 10 at each step. We can imagine this procedure going on indefinitely, with subdivisions that become ever finer; this lets us represent the number concerned with arbitrarily high precision. Real physics isn't like that at the atomic level, and space itself is probably not like that, but real numbers represent reality extremely well for many purposes.

*

Historically, new types of number generally met with resistance when first proposed. Then, as their utility became apparent, and their uses became established, people warmed to the idea. Within a generation, most resistance vanished: if you grow up using something regularly, it seems entirely natural. Philosophers could argue about whether zero is a number, and they still do, but ordinary folk used it when needed and stopped wondering what it was. Even mathematicians did that, though with occasional feelings of guilt. The terminology rather gives the game away: the new numbers are negative, irrational.

However, even among mathematicians, some innovations caused headaches, which lasted for centuries. What *really* set the cat among the pigeons was the introduction of so-called 'imaginary' numbers. Even the name (still in use for historical reasons only) suggests a degree of bewilderment, a hint that these numbers were somehow disreputable. Again, the underlying issue was square roots.

Once we've enlarged the number system to include infinite decimals, every positive number has a square root. In fact, it has two: one positive, the other negative. For example, 25 has two square roots, +5 and −5. This curious fact is a consequence of the rule 'minus times minus equals plus', which often baffles people when they first meet it. Some never accept it. However, it's a simple consequence of the principle that negative numbers ought to obey the same arithmetical rules as positive ones. That sounds reasonable, but it implies that *negative numbers don't have square roots*. For instance, −25 has no square root. This seems unfair, given that its cousin +25 has two of the things. So mathematicians speculated about a new realm of numbers, in which negative numbers *do* have square roots. They also tacitly assumed that in this expanded realm the usual rules of arithmetic and algebra continue to hold. It then became apparent that only one radically new number is needed: a square root of minus one. Giving this newfangled thing the symbol i that everyone except engineers now uses (they use j), its key feature is that

$$i^2 = -1$$

Now fairness reigns, and every number, positive or negative, has two square roots.[48] Except for 0, because −0 = +0, but zero is often exceptional, so no one worries about that.[49]

The idea that a negative number might possess a meaningful square root can be traced back to the ancient Greek mathematician and engineer Heron of Alexandria, but the first steps towards making sense of that idea were taken a millennium and a half later in Renaissance Italy. Girolamo Cardano mentioned the possibility in his *Ars Magna* (The Great Art, one of the first algebra texts) in 1545, but dismissed the idea as being pointless. A breakthrough came in 1572, when the Italian algebraist Rafael Bombelli wrote down rules for performing calculations with a hypothetical square root of minus one, and found the *real*-number solutions of a cubic equation using a formula that added together two 'numbers' that couldn't possibly be real. The impossible bits conveniently cancelled each other out, leaving the correct – and real – answer. This audacious piece of arcane trickery made mathematicians sit up and take notice, because those solutions could be checked directly, and they worked.

To sugar the pill, the new numbers were said to be 'imaginary', as opposed to traditional 'real' numbers that could be used to measure real objects. This terminology gave real numbers an undeservedly special status, and it confused a mathematical concept with a standard way to use it. As we'll see, imaginary numbers have perfectly sensible uses and interpretations, but not as measurements of standard physical quantities like length or mass. Bombelli was the first person to demonstrate that imaginary numbers – whatever the confounded things might be – could be used to solve entirely real problems. It was as if some weird carpenter's tool, that didn't even exist, could somehow be picked up and used to make a perfectly normal chair. Of course, it was a conceptual tool, but even so, the procedure was baffling. Even more baffling was the evidence that it worked.

Miraculously, it kept on working, in an ever-widening range of applications. By the 1700s mathematicians were making free use of these new numbers. Euler introduced the standard symbol i for the square root of minus one in 1777. Combining real and imaginary numbers led to a beautiful, self-consistent system known as the complex numbers – complex in the sense of 'composed of several parts', not 'complicated'. Algebraically they look like $a + b$i, where a and b are real numbers. You can add them, subtract, multiply, divide, take square roots, cube roots, and so on, without ever leaving the system of complex numbers.

The main deficiency is that it's hard to find an interpretation in the real world, or, at least, that's what everyone thought at the time. It's not clear what a measurement of, say, 3 + 2i, looks like. Quasi-philosophical debates about the legitimacy of complex numbers raged, until mathematicians discovered how to use them to solve problems in mathematical physics. Since the answers could be checked by other means, and always seemed to be correct, the debate went on the back burner in the rush to exploit these powerful new techniques.

*

For a long time, mathematicians attempted to justify imaginary numbers by appealing to a sweeping but vague 'principle

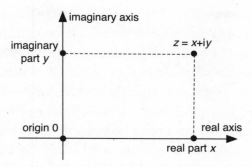

The complex number plane.

of permanence', which basically claimed that any algebraic rule that's valid for real numbers must automatically hold for complex ones as well. In a triumph of hope over logic, the main evidence for this claim was that, in practice, using complex numbers gave correct answers. In short, they worked because they worked, and as proof ... they worked.

Only much later did mathematicians sort out how to represent complex numbers. In fact, just like negative numbers, they have several different 'real-world' interpretations. We'll shortly see that in electrical engineering, a complex number combines the amplitude (maximum size) of an oscillating signal with its phase, in a single compact and convenient package. The same happens in quantum mechanics. More prosaically, just as real numbers model points on a line, complex numbers model points on a plane. It's that simple. And, like many simple ideas, it was overlooked for centuries.

The first hint of this breakthrough can be seen in the 1685 *Algebra* of John Wallis. He extended the standard representation of the real numbers as a line to complex numbers. Suppose the number is $a + b$i. The 'real part' a is just a standard real number, so we can locate it on the usual real line, which can be thought of as some fixed line in a plane. The remaining component bi is an imaginary number, so it doesn't match any point on that line. However, the coefficient b is real, so we can draw a line of length b in that plane, at right angles to the real line. The point in the plane thus obtained represents $a + bi$. Today we immediately see that this represents that number as the point in the plane with coordinates (a, b), but at the time Wallis's proposal fell on stony ground. Historical credit commonly goes to Jean-Robert Argand, who published it in 1806, but he had been beaten to the punch by a little-known Danish surveyor, Caspar Wessel, in 1797. However, Wessel's paper was in Danish, and escaped notice until a French translation appeared a century later. Both of them gave Euclidean-style geometric constructions showing how to add and multiply any two complex numbers.

Finally, in 1837, the Irish mathematician William Rowan Hamilton explicitly pointed out that you can represent a complex number as a pair of real numbers – the coordinates of a point in the plane:

complex number = (first real number, second real number)

Then he rewrote the geometric constructions as two formulas for adding and multiplying these pairs. I'll show you them because they're quite simple and elegant:

$$(a, b) + (c, d) = (a + c, b + d)$$
$$(a, b) \cdot (c, d) = (ac - bd, ad + bc)$$

This may seem a trifle cryptic, but it does the job beautifully. Numbers of the form $(a, 0)$ behave just like real numbers, and the mysterious i is the pair $(0, 1)$ – this is Wallis's suggestion that imaginaries are at right angles to reals, written in coordinates. Hamilton's formulas tell us that

$$i^2 = (0, 1) \cdot (0, 1) = (-1, 0)$$

which we've already identified with the real number −1. Job done. Naturally, it then turned out that Gauss had mentioned the same idea in a letter to Wolfgang Bolyai in 1831, but hadn't published it.

What Gauss perhaps didn't fully appreciate, but Hamilton did, is that those two formulas also let you prove that complex numbers obey all the usual rules of algebra, previously associated with only the real numbers. Rules like the commutative law $xy = yx$ and the associative law $(xy)z = x(yz)$, which most of us take for granted when we're first introduced to algebra. To prove they also work for complex numbers, replace the symbols by pairs of real numbers, apply Hamilton's formulas, and check that both sides give the same pair using just the algebraic rules obeyed by real numbers. Easy peasy. Ironically, by the time Gauss and Hamilton had sorted out the underlying logic in terms of pairs of ordinary 'real' numbers, mathematicians had made so much use of complex numbers that they'd pretty much lost interest in giving them a specific logical meaning.

Paramount among those uses were questions in physics, such as magnetic and electric fields, gravity, and fluid flow. Remarkably, some basic equations of complex analysis (calculus with complex functions) were a precise match for standard equations of mathematical physics. So you could solve the physics equations by doing calculus with complex numbers. The main restriction was that complex numbers lie in a plane. So the physics also had to take place in a plane, or be equivalent to a problem in a plane.

*

Complex numbers give the plane a systematic algebraic structure that's beautifully adapted to geometry, hence also to motion. You can consider the rest of this section to be a two-dimensional dry run at similar issues in three-dimensional geometry, the subject of the next chapter. There will be a few formulas – it *is* algebra,

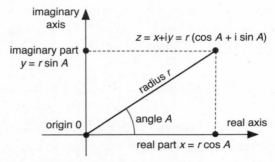

Cartesian and polar coordinate geometry of the complex plane.
Here cos and sin are the trigonometric functions cosine and
sine. (The picture effectively defines these functions.)

after all – but I'm not sure how to avoid them without everything seeming rather vague.

When we represent a complex number z in the form $z = x + iy$, with x and y real, the underlying geometry is the Cartesian coordinate system, named after René Descartes, with two sets of axes at right angles to each other: the real part x (horizontal axis) and the imaginary part y (vertical axis). However, there's another important coordinate system in the plane, called polar coordinates, which represents a point as a pair (r, A) where r is a positive real number and A is an angle. These two systems are closely related: r is the distance from the origin 0 to z, and A is the angle between the real axis and the line joining the origin to z.

Cartesian coordinates are ideal for describing how objects move without rotating. If a point $x + iy$ is displaced a units horizontally and b units vertically, it moves to $(x + iy) + (a + ib)$. If we extend this idea to a set of points, with a list of values for x and y, then the entire set moves a units horizontally and b units vertically if we add a fixed complex number $a + ib$ to each point in the set. Moreover, this motion is *rigid*: the entire object moves without changing shape or size.

Another important type of rigid motion is rotation. Again, the

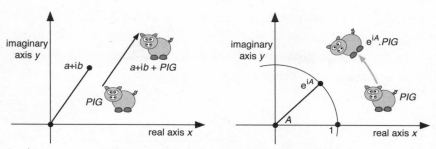

Translating (*left*) and rotating (*right*) the point
set *PIG* using complex numbers.

object doesn't change shape or size, but its orientation changes, and it spins through some angle about some central point. A key observation here is that multiplication by i rotates points through a right angle, about a centre at the origin. This is why the y-axis, which represents the 'imaginary part' y of z, is at right angles to the x-axis, which represents the 'real part' x. (Despite the name, the imaginary part is a real number: it *becomes* imaginary when we multiply by i to get iy.)

If we want to rotate a set of points through a right angle, we multiply every point in the set by i. More generally, if we want to rotate a set of points through some angle A, a bit of trigonometry reveals that we have to multiply them all by the complex number

$$\cos A + i \sin A$$

Euler discovered a remarkable and beautiful connection between this expression and the complex analogue of the exponential function e^x, where $e = 2\cdot71828\ldots$ is the base of natural logarithms. We can define the exponential function e^z of a complex number z so that it has the same basic properties as the real exponential, and agrees with it when z is real. It turns out that

$$e^{iA} = \cos A + i \sin A$$

One elegant way to see why this happens is to use differential equations: I've put it in the Notes[50] because it's slightly too technical.

In the polar coordinate representation of a complex number, the coordinates turn out to represent the point

$$r(\cos A + i \sin A) = re^{iA}$$

which is a simple, compact formula.

The beauty of the complex numbers, where geometry is concerned, is that they simultaneously have two natural coordinate systems, Cartesian and polar. Translating an object has a simple formula in Cartesian coordinates, but in polar coordinates it's a mess. Rotating an object has a simple formula in polar coordinates, but in Cartesian coordinates it's a mess. If you use complex numbers, you can choose which representation best suits your purposes.

These geometric features of complex algebra could be exploited in two-dimensional computer graphics, but it turns out that because the plane is simple and computers don't mind messy formulas, you don't gain a lot by doing so. In Chapter 7 we'll see that for computer graphics in three dimensions, a similar trick works marvels. For the moment, however, we need to polish off the story of complex numbers by discussing some genuinely useful applications.

*

Mathematicians gradually came to realise that despite the lack of an obvious physical interpretation, complex numbers are often simpler than real numbers, and shed light on features of real numbers that are otherwise puzzling. For instance, as Cardano and Bombelli noticed, quadratic equations have either two real solutions or none, and cubic equations have either one real solution or three. It's much simpler for complex solutions: quadratic equations always have two complex solutions, and cubic equations always have three. For that matter, tenth-degree equations have ten complex solutions, but they might have 10, 8, 6, 4, 2, or no real solutions. In 1799 Gauss proved a long-suspected fact, conjectured by Paul Roth as long ago as 1608, which became known as the Fundamental Theorem of Algebra: a polynomial equation of degree n has n complex solutions. All of the standard functions of analysis, such as the exponential, sine, cosine, and so on, have natural complex analogues, and their properties generally become simpler when we view them from a complex perspective.

One practical consequence is that complex numbers have become a standard technique in electronic engineering, mainly because they provide an elegant and simple way to deal with alternating currents. Electricity is a flow of electrons, charged subatomic particles. In direct current, which is produced for example by a battery, all the electrons flow in the same direction. In alternating current, widely used in mains electricity because it's safer, the electrons shuttle back and forth. The graph of voltage (and current) looks like a cosine curve in trigonometry.

A simple way to produce this curve shows up if you think about a point on the rim of a rotating wheel. Suppose the wheel has radius 1, for simplicity. If you look at the horizontal projection of the rotating point, it moves from side to side, reaching values $+1$ and -1 at its extremes. If the wheel rotates at a constant speed, the graph of this horizontal distance is a cosine curve, and the graph of the vertical distance is a sine curve (solid curves in the picture).

The position of the moving point is the pair of real numbers ($\cos A$, $\sin A$) where is A is the angle between the point and the horizontal axis. Using Hamilton's trick we can interpret this as the complex number $\cos A + i \sin A$. As A varies, this number moves round and round the unit circle in the complex plane. If we measure angles in radians, it makes a complete circle when A increases from 0 to 2π. Then it makes another one when A

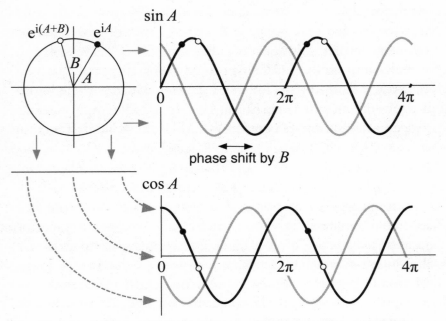

Rotation in the complex plane projects to give periodic oscillations.
Adding B to angle A moves the graphs to the left: a phase shift.

increases from 2π to 4π, and so on, so the motion is periodic
with period 2π.

Euler's formula implies that as A runs through the real
numbers, the corresponding value of e^{iA} goes round and round the
unit circle at a constant speed. This connection provides a way to
turn any statement about an oscillating function shaped like the
sine or cosine into a complex exponential. Mathematically, the
exponential is simpler and more tractable. Moreover, the angle
A has a natural physical interpretation as the phase of the oscillation, which means that changing A by adding a constant angle
B shifts the sine and cosine curves by the corresponding amount
(grey curves in picture).

Even better, the basic differential equations for voltages
and currents in circuits extend unchanged to the corresponding

complex equations. The physical oscillation becomes the real part of a complex exponential, and the same methods apply to alternating current as well as to direct current. It's as though the real behaviour has a secret imaginary companion, and the two together are simpler than either on its own. Electronic engineers routinely use this mathematical trick to simplify their calculations, even when using a computer.

*

In this application to electronics, the complex numbers are plucked like a mathematical rabbit from a conjurer's hat, which just happens to make life simpler for engineers. But there's a remarkable context in which complex numbers are absolutely necessary, and have a physical meaning. Namely, quantum mechanics.

Wigner made this example of unreasonable effectiveness the centrepiece of his lecture:

> Let us not forget that the Hilbert space of quantum mechanics is the complex Hilbert space ... Surely to the unpreoccupied mind, complex numbers are far from natural or simple and they cannot be suggested by physical observations. Furthermore, the use of complex numbers is in this case not a calculational trick of applied mathematics but comes close to being a necessity in the formulation of the laws of quantum mechanics.

He also went out of his way to emphasise what he meant by 'unreasonable':

> Nothing in our experience suggests the introduction of these quantities. Indeed, if a mathematician is asked to justify his interest in complex numbers, he will point, with some indignation, to the many beautiful theorems in the

theory of equations, of power series and of analytic functions in general, which owe their origin to the introduction of complex numbers ... It is difficult to avoid the impression that a miracle confronts us here, quite comparable ... to the two miracles of the existence of laws of nature and of the human mind's capacity to divine them.

Quantum mechanics came into being around 1900 to explain the strange behaviour of matter on small scales that experimental physicists had started to discover, and it rapidly grew into the most successful physical theory that humanity has yet invented. Down on the level of molecules, atoms, and especially the subatomic particles that come together to create atoms, matter behaves in surprising and baffling ways. So surprising and baffling that it's not at all clear that the word 'matter' applies. Waves, such as light, sometimes behave like particles: photons. Particles, such as electrons, sometimes behave like waves.

This wave–particle duality was eventually resolved with the introduction of mathematical equations that govern both waves and particles, although even now much remains puzzling. In the process, the way both are represented in the mathematics underwent, as Shakespeare put it, 'a sea-change. Into something rich and strange'. Up to that time, physicists characterised the state of a particle of matter in terms of a small list of numbers: mass, size, position, velocity, electric charge, and so on. In quantum mechanics the state of any system is characterised by a wave; more precisely, its wave function. As the name suggests, this is a mathematical function with wavelike properties.

A function is a mathematical rule or process that transforms some number into another number in a specified manner. More generally, a function can transform a list of numbers into a number, or even another list of numbers. More generally still, a function can operate not just on numbers, but on sets of mathematical objects of any kind. For example, the function 'area' acts on the set of all triangles, and when you apply it to any given triangle, the output of the function is that triangle's area.

The wave function of a quantum system acts on the list of possible measurements that we might make on the system, such as its position coordinates or its velocity coordinates. In classical mechanics, finitely many such numbers normally determine the state of the system, but in quantum mechanics this list might involve infinitely many variables. These are taken from a so-called Hilbert space, which is (often) an infinite-dimensional space with a well-defined notion of the distance between any two of its members.[51] The wave function outputs a single number for each function in the Hilbert space, but the number that it outputs isn't a real number: it's a complex one.

In classical mechanics, an observable (a quantity we can measure) associates to each possible state of the system a number. For instance, when we observe the distance from the Earth to the Moon we come up with a single number, and this is a function defined on the space of all possible configurations that the Earth and Moon might in principle take up. In quantum mechanics, observables are *operators*. An operator takes an element of the Hilbert space of states and turns it into a complex number. Operators have to obey a short list of mathematical rules. One is linearity. Suppose you have two states x and y, and the operator L outputs $L(x)$ and $L(y)$ for these. In quantum theory, states can superpose – add together – to give $x + y$. Linearity means that the operator L must then give the output $L(x) + L(y)$. The full list of required properties gives a so-called Hermitian operator, which behaves nicely in connection with distances in the Hilbert space.

Physicists choose these spaces and operators in various ways to model specific quantum systems. If they're interested in the states of position and momentum for a single particle, the Hilbert space consists of all 'square-integrable' functions, which is infinite-dimensional. If they're interested in the spin of a single electron, the Hilbert space is two-dimensional, consisting of so-called 'spinors'. An example is Schrödinger's equation, which looks like this:

$$i\hbar \frac{\mathrm{d}}{\mathrm{d}t}|\Psi(t)\rangle = \hat{H}|(t)\rangle$$

You don't need to understand the mathematics, but let's look at the symbols. Especially the first one, which gives much of the game away: it's i, the square root of minus one. We're looking at the basic equation of quantum mechanics, and the first symbol we see is the imaginary number i.

The next, $\hbar$, is a number called the reduced Planck's constant, and it's very, very tiny: about 10^{-34} joule seconds. It's what gives quantum mechanics its quanta – tiny but discontinuous jumps in the sizes that various quantities can assume. Then there's a d/dt fraction. The t is time, and the d's tell us to find a rate of change, as in calculus, so it's a differential equation. The combination of symbols $|\Psi(t)\rangle$ is the wave function, which specifies the quantum state of the system at time t, so this is the thing whose rate of change we want to know. Finally, $\hat{H}$ is the so-called Hamiltonian: basically, the energy.

The usual interpretation of the wave function is that it represents not any individual state, but the *probability* that an observation will find the system in that state. However, probabilities are real numbers between 0 and 1, whereas the outputs of the wave function are complex numbers of any size. So physicists focus on the amplitude (which mathematicians call the modulus) of the complex number, which is how far it is from the origin – the r in polar coordinates. They think of that number as a relative probability, so if one state has amplitude 10 and another has amplitude 20, then the second is twice as probable as the first.

The modulus tells you how far away from the origin a complex number lies, but it doesn't tell you the direction in which you have to go to reach it. This direction is specified by another real number, the angle A in polar coordinates. Mathematicians call this angle the argument of the complex number, but physicists call it the phase – how far round the unit circle you need to go. So the complex wave function has an amplitude, which quantifies the relative probability of that observation occurring, and a phase, which doesn't alter the amplitude and is almost impossible to measure. Phases affect how states superpose, hence also the probabilities of those superposed states occurring, but in practice they're hidden from experimental sight.

What all this means is that a real number alone is the wrong kind of number to quantify a quantum state. You can't even *formulate* quantum mechanics in terms of conventional real numbers.

*

If the question is 'What practical uses do complex numbers have?', then we can point to all of the myriad applications of quantum mechanics, safe in the knowledge that they must also be applications of complex numbers. Until recently, most of the answers would have been about laboratory experiments – deep frontier physics, but not the kind of thing you find in your kitchen or living room. Modern electronics has changed all that, and many of our favourite devices function for quantum-mechanical reasons. The engineers need to understand such things in great depth and detail, but we can just sit back and admire their creations. Or, from time to time, curse them when they fail to do what we want because of some obscure technicality in how we've configured the confounded thing.

My newly installed fibre broadband is a case in point. It looks like conventional cable, but it's part of a transmission system that already relies on quantum technology. The quantum bit isn't in the cable as such, though: it's in the devices along the way that create the light pulses on which the whole set-up relies. Of course, light is really quantum anyway, but these devices are *designed* using quantum mechanics, and wouldn't work without it.

The word 'fibre' refers to a many-stranded cable, whose individual strands are thin glass threads that transmit light. They're designed so that the light bounces off the walls instead of escaping, so you can bend the cables round corners and the light stays inside the cable. Information is encoded in the light beam as a series of sharp pulses. The telecoms industry introduced optical fibre because it combined several advantages. The fibres now available are highly transparent, so they transmit the light over long distances without degrading the signal. Light pulses can carry far more information than traditional copper telephone wires. This higher bandwidth is what gives the increased 'speed' – it's not so much how fast the pulses move, it's how many pulses, how much information, can be crammed into one fibre or one cable. Fibre cables are lighter than copper ones, so they're easier to transport and install, and less prone to electrical interference.

An optical communications system has four main components: a transmitter (light source); a cable to carry the signal; a series of repeaters that pick the signal up before it has degraded too much, clean it up, and send it on; and, of course, a receiver (detector). I'll focus on just one, the transmitter. This has to be a device that can create light, and be controlled so that the light emerges as a series of single pulses, which can be switched on (1) or off (0) to encode a message in binary. The switching has to be extremely rapid, and everything has to be very precise. In particular, the wavelength ('colour') of the light should have one specific value. Finally, the pulses need to retain their form, so that the receiver can recognise them.

The ideal (indeed, the only) gadget for this is the laser, a device that emits a powerful beam of coherent light of some specific wavelength. 'Coherent' means that all of the waves in the beam are in phase with each other, so they don't cancel each other out. A laser does this by bouncing light (in the form of photons) to and fro between a pair of mirrors, triggering an ever-increasing cascade of photons in a positive feedback loop. When the beam becomes strong enough, it's allowed to escape.

The first lasers were large and cumbersome, but today most of the lightweight lasers are made by the same general processes that produce the microscopic circuits in computer 'chips' – integrated semiconductor circuits. For the last thirty years, almost all lasers used in consumer and corporate technology (such as blu-ray players, which were made possible by lasers that produce blue light) have been Separate Confinement Heterostructure (SCH) lasers. These are a refinement of quantum well lasers, which are a kind of sandwich whose middle layer acts as a quantum well. This creates wave functions looking like a series of steps, rather than a curve, so the energy levels are quantised – sharp and separate rather than fuzzy and merging together. These levels can be tuned, by suitable design of the quantum well, to create light of the right frequency for laser action.

SCH lasers add two further layers to the top and bottom of the sandwich, with lower refractive index than the middle three, which confines the light within the laser cavity. It stands to reason that you can't work out how to engineer a quantum device of this kind without applying a lot of quantum mechanics. So even the fibre optics of the 1990s was already using quantum components, and the same is even truer today.

In future, a huge range of novel quantum devices is likely to transform our lives. The Heisenberg uncertainty principle for quantum mechanics tells us that certain observables can't be measured exactly at the same time – for instance, if you know exactly where a particle is, you can't be sure how fast it's moving. This feature can be used to detect whether an unauthorised person is listening in to secret messages. When Eve the eavesdropper sneakily observes the quantum state of a passing signal – say the spin of a photon – that state changes, and she can't control how it changes. It's like a bell built into the message that rings whenever Eve tries to read it.

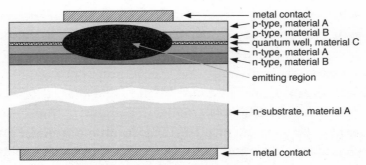

Schematic structure of SCH laser. The terms n-type and p-type refer, respectively, to semiconductors where the charge is carried by electrons or by 'holes', where electrons are missing.

One way to implement this idea is to use quantum photonics, the quantum mechanical properties of photons. Another is to manipulate the spins of quantum particles: the developing field of spintronics. Such devices can convey more information than a conventional signal by encoding extra data in the spins of the particles, not just whether they're present. So my Superfast Fibre Broadband may soon be superseded by SuperDuperFast Spintronic Broadband, carrying far more information along the same cable. Until some bright spark invents Six-Dimensional Hyper-Definition Sensory Holography and gobbles up all that extra bandwidth.

7

Papa, Can You Multiply Triplets?

Ocean waves crashing nicely against your boat in *ASC: Black Flag*? Math.
Those bullets flying over your head in *Call of Duty: Ghosts*? Math.
Sonic being able to run fast and Mario being able to jump? Math.
Drifting around that corner in *Need for Speed* at 80 mph? Math.
Snowboarding down a slope in *SSX*? Math.
That rocket blasting off in *Kerbal Space Program*? Math.

> Forbes website, 'This is the Math Behind Super Mario'

The village has a medieval look, with thatched cottages, horse-drawn carts on a dirt road, fields with crops and sheep. The narrow ribbon of a river flows between the close-packed buildings, glinting golden in the setting sun. We see the scene from above, as if viewed from an aircraft; the view spins and sways as the aircraft swoops and twists. But this is no aircraft: a cut to a viewpoint on the ground shows the clear outline of a dragon. Incoming. Cut back to the dragon's-eye view, now in a steep dive, skimming the rooftops, a stream of flame gushing in front of it as thatch catches fire ...

It might be a movie, it might be a computer game – nowadays they can be almost indistinguishable. Either way it's a triumph of computer graphic imagery – CGI.

Is this mathematics?

Oh, yes.

It must be very new, then.

Not exactly. The *application* is new, and some of the mathematics is both new and sophisticated, but the part of it that I have in mind is about 175 years old. And that bit of the mathematics was never intended for computer graphics. There weren't any computers in those days.

It *was* intended to handle a more general issue, independent of any hardware: geometry in three-dimensional space. From today's perspective, the potential for a connection to computer graphics is clear. But it didn't look like geometry. It looked like algebra. Except that it broke one of the basic algebraic rules, so it didn't even look like algebra. It was conjured into existence by the Irish mathematical prodigy Sir William Rowan Hamilton, who called his brainchild *quaternions*. Ironically, quaternions weren't exactly what he'd been looking for, and there was a reason for that.

What he'd been looking for didn't exist.

*

Today there are more computers on the planet than there are people. The number of human beings is over 7·6 billion. There are more than 2 billion laptops alone, and nearly 9 billion smartphones and tablets, both of which often have more computing power than the best supercomputer money could buy in 1980.[52] Counting the tiny computers that manufacturers are rushing to shoehorn inside every dishwasher, toaster, refrigerator, washing machine, and catflap on the planet, computers now outnumber humanity four to one.

It's hard to realise that it wasn't always this way. The pace of innovation has been explosive. The first home computers – Apple II, TRS-80, Commodore PET – hit the consumer market in 1977, just over forty years ago. Almost from the start, a major use of home computers was to play games. The graphics were clunky, the games very simple. Some consisted only of text messages: 'You are in a maze of twisty passages, all different.' Followed by an even more sinister message: 'You are in a maze of twisty passages, all alike.'

As computers became faster, memory became almost infinite, and prices plummeted, computer-generated images became far more convincing, so much so that they began to take over the movie industry. The first feature-length animated movie produced entirely by computer was *Toy Story* in 1995, though shorter examples go back another decade. Special effects have now become ultra-realistic, and are so widely used that we hardly notice they're there. When Peter Jackson filmed the *Lord of the Rings* trilogy, he didn't worry about the lighting: it was sorted out afterwards, post-processed by computers.

We have become so used to high-quality, fast-moving graphics that we seldom stop to wonder where it all came from. When did the first video game appear? Thirty years before home computers. In 1947 the TV pioneers Thomas Goldsmith Jr and Estle Ray Mann filed a patent for a 'cathode-ray tube amusement device'. A cathode-ray tube is a short fat glass bottle with a wide, gently curving base – the screen – and a narrow neck. A device in the neck fires a beam of electrons at the screen, and electromagnets control the direction of the beam, scanning it across the screen in a series of horizontal sweeps, like a human eye reading a page of text. Where the beam hits the front of the tube, it causes a special coating to fluoresce, creating a bright spot of light. Most TV sets used cathode-ray tubes for their display until flat-screen TVs appeared commercially around 1997. Goldsmith and Mann's game was inspired by Second World War radar displays. The spot of light represented a missile, and the player tried to make it hit targets, which were drawn on paper and stuck to the screen.

By 1952 the mainframe computer EDSAC had scaled the giddy heights of playing noughts and crosses. The big consumer hit was

Pong, an early arcade game manufactured by Atari – a simplified two-dimensional game of table tennis with a ball that bounced off two paddles, one for each player. By today's standards the graphics were very basic, two moving rectangles for the paddles and a moving square for the ball, and the action was almost non-existent, but until better technology became available *Pong* was state-of-the-art video gaming.

Needless to say, Hamilton could not have intended his mathematical brainchild to be used in such a manner. That idea took another 142 years to germinate. But we can see, with hindsight, that this possibility was inherent in the type of problem that his discovery was intended to help solve. There are many styles of mathematics. Mathematicians can be problem solvers, focused on finding the answer to a specific question, be it in the real world or the mental world of pure mathematics. They can be theory builders, organising innumerable special theorems within a unified framework. They can be mavericks, roaming erratically from one area to the next, working on whatever takes their fancy. Or they can be toolmakers, fashioning new tools that might be useful when tackling questions that have not yet been posed – a method in search of an application.

The bulk of Hamilton's reputation rests on his work as a theory builder, but quaternions illustrate his prowess as a toolmaker. He invented them to provide an algebraic structure for systematic calculations about the geometry of three-dimensional space.

*

Hamilton was born in Dublin, Ireland, in 1805, the fourth child out of nine. His mother was Sarah Hutton and his father was Archibald Hamilton, a solicitor. When William was three years old he was sent to live with his uncle James, who ran a school. William had a precocious talent for languages, but he seems to have taught himself a great deal of mathematics as well, and that was the subject he studied at Trinity College Dublin from the age of 18, getting sky-high marks. John Brinkley, Bishop of Cloyne, declared that 'This young man, I do not say *will be*, but *is*, the first mathematician of his age.' The bishop was arguably right, and in 1837, while Hamilton was still an undergraduate, he was made the Andrews Professor of Astronomy and Royal Astronomer of Ireland. He spent the remainder of his professional life at Dunsink Observatory near Dublin.

His most celebrated works were in optics and dynamics, especially the discovery of a remarkable link between these two diverse areas of mathematical physics, which Hamilton reformulated in terms of a common mathematical concept, the principal function. We now call it the Hamiltonian, and it led to major advances in both areas. Later it turned out to be just what was needed for the newfangled and very strange theory of quantum mechanics.

We met Hamilton briefly in the previous chapter. In 1833 he resolved a centuries-old quasi-philosophical conundrum that stripped complex numbers of their mystery, revealing them to be impostors, their apparent novelty resulting from a cunning disguise, their true nature almost trivial. A complex number, said Hamilton, is nothing more nor less than an ordered pair of real numbers, equipped with a specific list of rules for adding and multiplying pairs. We also saw that this resolution of the puzzle came just too late to impress anybody, and that when Gauss had come up with the same idea he hadn't even bothered to publish it. Nonetheless, Hamilton's way of thinking about complex numbers proved of great value, because it inspired him to create the quaternions.

For these mathematical advances, and others, Hamilton was knighted in 1835. Quaternions came later, and when they did, few people, other than Hamilton himself and a few devotees, appreciated their importance. I think that during his lifetime most mathematicians and physicists viewed his enthusiastic promotion of quaternions as some kind of bee in his bonnet – not exactly crackpottery, but perilously close to it. They were wrong. His new invention triggered a revolution, leading mathematics into wild, uncharted territory. You can see why most failed to appreciate its potential, but Hamilton knew he was on to something. His wild territory is still offering tantalising new insights today.

*

Questions that very few gamers or moviegoers worry about are: How do the graphics work? How are these illusions created? What makes them so convincing? Fair enough: you don't need to know about any of that to enjoy playing the game or watching the movie. But the historical development, the techniques that had to be invented to make it possible, and the companies that specialise in CGI and write games, need lots of highly trained people who know how the various tricks work, in considerable technical detail, and have the mastery and creativity to invent new ones. It's not an industry in which you can rest on your laurels.

The basic geometric principles have been around for at least 600 years. During the Italian Renaissance, several prominent painters began to understand the geometry of perspective drawing. These techniques let the artist create realistic images of a three-dimensional world on a two-dimensional canvas. The human eye does much the same thing, with the retina in place of the canvas. A full description is complicated, but in simple terms, the artist *projects* a real scene onto a flat canvas by constructing a straight line from each point of the scene to a point representing the viewer's eye, and marking the place where this line meets the canvas. Albrecht Dürer's wonderful woodcut *Man Drawing a Lute* is a vivid depiction of this procedure.

This geometric description can be turned into a simple mathematical formula, which transforms the three coordinates of a point in space to the two coordinates of the corresponding image

Albrecht Dürer's *Man Drawing a Lute*, illustrating a projection
from three-dimensional space to a two-dimensional canvas.

on the canvas. To apply the formula, you just have to know the
positions of the canvas and the viewer's eye relative to the scene.
For practical reasons you don't apply this transformation, called
a projection, to *every* point of the object, but to enough points to
give a good approximation. This feature is visible in the woodcut,
which shows a lute-shaped set of dots, not the full outlines of the
lute. Fine details, such as the thatch of a roof, the ripples on the
river, and of course their colours, can then be 'draped' over this
collection of points, using methods that I won't go into because
we'd need another book.

This is basically what happens when we're shown a dragon's-
eye view of the village. The computer already has representative
coordinates of every important feature of the village stored in
memory. The dragon's retina plays the role of the canvas. If we

know where it is, and at which angle, we can use the formula to calculate what the dragon would see. That gives one frame of the movie, showing the village at a specific instant of time. In the next frame, the village is still in the same place, but the dragon, and its retina, have moved. Work out where they've gone, repeat the calculation, and you have the next frame. Follow the dragon's path through the sky, and frame by frame you put together the movie of what it sees.

This isn't a literal description, of course; just the main underlying idea. There are special tricks to make the calculations more efficient, saving processing time on the computer. For simplicity, let's ignore those.

The same type of calculation applies for scenes of the inbound dragon, viewed from the ground. Now we need another set of points to specify where the dragon is, and the screen onto which everything is projected is on the ground, not the dragon. For definiteness, let's settle on the dragon's view. From her viewpoint, her eye is fixed, and it's the *village* that seems to move. As she swoops towards the ground, everything in the village seems to get bigger, and it tilts and turns, mimicking her own movements. As she soars towards the clouds, the village shrinks. Throughout, the perspective has to remain convincing, and the mathematical key to that is to treat the village as a *rigid* (and quite complex) object. You can get an idea of what's involved by pretending to be the dragon, holding some object in front of your eyes, and then moving it to and fro, spinning it this way and that.

Now we're representing everything in the dragon's 'frame of reference', which is fixed *relative to the dragon*. The village moves like a rigid body, which mathematically means that the distance between any two points remains the same. But the body as a whole can move in space. There are two basic types of movement: translation and rotation. In translation, the body slides in some direction without twisting or turning. In rotation, the body spins about some fixed line, the axis, and every point moves through the same angle in a plane cutting the axis at right angles. The axis can be any line in space, and the angle can have any size.

Every rigid motion is a combination of a translation and a rotation (but the translation can be through a zero distance, and the rotation might be though a zero angle, in which case those transformations have no effect). Actually, that's a lie: another possible rigid motion is a reflection, which acts like a mirror. But you can't get reflections by continuous movement, so we can ignore those.

We've now made the key step in turning moving dragons into mathematics. What we need to understand is how the coordinates of a point in space change when we apply a translation or a rotation. Having done that, we can use the standard formula to project the result onto a flat screen. It turns out that translations are easy. The big bugbear is rotations.

*

Everything is much easier in two dimensions – a plane. Euclid formalised the geometry of the plane around 300 BC. However, he didn't set it up using rigid motions. Instead, he used congruent triangles – triangles that are the same shape and size, but in different positions. By the nineteenth century, mathematicians had learned to interpret such a pair of triangles as a rigid motion, a transformation of the plane that moves the first triangle to the position of the second. Georg Bernhard Riemann *defined* geometry in terms of specific types of transformation.

Following a very different route, mathematicians had also come up with efficient ways to calculate rigid motions in a plane, as an unexpected side effect of a new development in algebra which we met in the previous chapter: complex numbers. To translate (slide) a shape, such as the *PIG* pictured on page 139, we add a fixed complex number to every point of that shape. To rotate it through angle A, we multiply every point by e^{iA}. As the icing on the cake, complex numbers were ideal for solving the differential equations of physics … but only in two-dimensional space.

All this gave Hamilton an idea, which became an obsession. Since complex numbers are so effective for physics in two dimensions, there ought to be analogous 'supercomplex' numbers that do the same for three dimensions. If he could find a new system of numbers like that, the whole of realistic physics would be wide open. It was even obvious how to get started. Since complex numbers are *pairs* of real numbers, these hypothetical supercomplex numbers ought to be *triples* of real numbers. One real per dimension. The formula for adding such triples (or triplets, as Hamilton often called them) was obvious – just add corresponding components. Translations sorted. All he had to do now was to find out how to multiply them. But everything he tried failed, and by 1842 he was so obsessed with this obstacle that even his children had noticed. Every day they would ask: 'Papa, can you multiply triplets?' And every day, Hamilton would give a wry shake of the head. Add or subtract them, yes: multiply them – no way.

It's often difficult to work out the exact date at which a great mathematical breakthrough occurred, because there's often a lengthy and confused 'prehistory' during which mathematicians were groping towards the eventual discovery. But sometimes we know the exact time and place. Here, the crucial date is Monday 16 October 1843, and the place is Dublin. We can even make a good stab at the time, because Hamilton, by then the President of the Royal Irish Academy, was walking along a canal towpath with his wife on his way to a meeting of the Academy's Council. As he took a breather on Brougham Bridge, the solution to the problem that had been vexing him for years flashed across his mind, and he carved it into the stonework with his pocket-knife:

$$i^2 = j^2 = k^2 = ijk = -1$$

The inscription has since worn away, but every year a group of scientists and mathematicians stride out along the 'Hamilton Walk' to keep the memory alive.

Without explanation, this inscription is hopelessly obscure. Even *with* explanation, it may well seem outlandish and pointless, on first encounter, but that's often the case with the great mathematical breakthroughs. They take time to sink in. If the discovery had been complex numbers, Hamilton would have carved a simple rule: $i^2 = -1$. This equation holds the key to the entire complex number system; all else follows if you insist that the usual rules of arithmetic continue to apply. Throw in j and k, as well as i, and Hamilton's formulas define a more extensive system of numbers – if you prefer, number-like objects. He named them *quaternions* because they have four components, each a conventional real number. These components are an ordinary real number, a real multiple of a number called i, which behaves just like the usual imaginary number with that symbol, and two new components: a real multiple of a number called j, and a real multiple of a number called k. A typical quaternion is thus a combination $a + bi + cj + dk$, where a, b, c, d are four ordinary real numbers. Or, to remove any mystery, a quadruple (a, b, c, d) of real numbers, obeying a short list of arithmetical rules.

The day after his minor act of vandalism, Hamilton wrote to his friend, the mathematician John Graves: 'There dawned on me the notion that we must admit, in some sense, *a fourth dimension* of space for the purpose of calculating with triples.' In a letter to his father, he wrote: 'An electric circuit seemed to close, and a spark flashed forth.' He spoke more truly than he could have known, because today his discovery plays a vital role in billions of electric circuits implementing quadrillions of tiny sparks. They go by such names as Playstation 4, Nintendo Switch, and Xbox, and they're used to play video games such as *Minecraft*, *Grand Theft Auto*, and *Call of Duty*.

We now understand why Hamilton had experienced so much

trouble trying to multiply triplets. It can't be done. He'd been assuming that all the usual laws of algebra must apply, and in particular that you can divide by any nonzero number, but no matter which formula he tried, it failed to obey all of the necessary laws. Later algebraists proved that this requirement is logically contradictory. If you want all of the laws to hold, you can't go beyond complex numbers. You're stuck in two dimensions. If you play about with Hamilton's formulas, and assume that the associative law holds, you'll quickly find that he'd already discarded one such law, the commutative law of multiplication. For example, his formulas imply that $ij = k$, whereas $ji = -k$.

Hamilton had the imagination to abandon this law, even though this was troublesome, to say the least. But we now know that even then you still can't construct a self-contained number system of triples. A beautiful theorem of Adolf Hurwitz, published posthumously in 1923, tells us that the real numbers, complex numbers, and quaternions are the only 'real division algebras'. That is, you can do the trick with one, two, or four real components, *but not three*. Of these, only the real and complex numbers obey the commutative law. By weakening the associative law you can also get a system with eight components, called the octonions or Cayley numbers. The next natural number of components would be 16, but now even the weakened form of associativity fails. That's it. Nothing else along those lines is possible. It's one of those weird curiosities that mathematics sometimes serves up: in this context, the next term in the sequence 1, 2, 4, 8, ... doesn't exist.

So poor old Sir William spent years of fruitless effort trying to achieve the impossible. His eventual breakthrough relied on abandoning *two* key principles: that multiplication ought to be commutative, and that the 'right' number system for three-dimensional physics should have three components. He deserves huge credit for realising that to make progress, you had to abandon both.

*

Hamilton's name for his new system, quaternions, reflects their relation to four dimensions. He promoted their use in many areas of mathematics and physics, showing that a special type of quaternion, the 'vector part' $b\mathrm{i} + c\mathrm{j} + d\mathrm{k}$, can represent three-dimensional space in an elegant manner. However, quaternions went out of fashion when a simpler set-up, vector algebra, came into being. They remained of interest in pure mathematics and theoretical physics, but failed to live up to their creator's hopes of practical uses. Until, that is, computer games and CGI in the movie industry came along.

The link to quaternions arises because CGI objects have to be rotated in three-dimensional space. The best method for doing this relies on Hamilton's quaternions. These provide a simple algebraic tool to calculate the effects of rotations quickly and accurately. Hamilton would have been amazed, because movies didn't exist in his time. Old mathematics can acquire radically new uses.

The proposal to use quaternions in computer graphics appears in a 1985 paper by Ken Shoemake, 'Animating rotation with quaternion curves'.[53] The paper opens with the statement 'Solid bodies roll and tumble through space. In computer animation, so do cameras. The rotations of these objects are best described using a four-coordinate system, quaternions.' Shoemake continues by stating that quaternions have the key advantage of allowing smooth 'in-betweening'; that is, interpolation of images between two given endpoints.

Before going into details, it's worth discussing a few features of computer animation that motivate his approach. This discussion is greatly simplified and many other techniques are also used. A movie or a moving image on a computer screen is actually a series of still images, shown in rapid succession to create the illusion of movement. In the early days of animation – think Walt Disney cartoons – artists drew each of these still images as a single piece of artwork. It took great skill to create realistic movements (inasmuch as a talking mouse can be realistic). Various tricks could be used to simplify the process, such as having a single background that remained the same throughout a sequence, and superimposing the objects that changed.

This method is very laborious, and impractical for rapid-action space battles or any other high-quality animation. Imagine you're animating a sequence from a movie or game in which several spacecraft interact. Each craft has already been designed (on a computer) by a graphic artist. It's represented as a fixed collection of points in space, linked together to form a network of tiny triangles. These in turn can be represented by suitable lists of numbers – coordinates of the points, and which connect to which. Computer software can 'render' this collection of numbers (and others such as colour) to create a two-dimensional image of the spaceship. This shows what it would look like when placed in some reference position and viewed from some particular location.

To make the spaceship move, the animator changes those numbers in an appropriate manner. For example, to move it to a new location, a fixed triple of numbers (the displacement vector) is added to all the points, while the links remain as before. Then this new list is rendered to get the next still image, and so on. Adding a vector is simple and quick, but objects can also rotate in space. They can rotate about any axis, and that axis may change as the object moves. Rotations also change the lists of numbers, but in more complicated ways.

Quite often, the animator knows where the object starts from (on the ground, say) and where it has to move to (lined up facing the distant Moon). The precise position on the two-dimensional screen is vital, because this is what the viewer sees. It has to look suitably artistic or exciting. So these two locations, start and finish, are represented by two carefully calculated lists of numbers. If the precise motion in between is less crucial, the computer can be instructed to interpolate between start and finish. That is, the two lists are combined together by a mathematical rule that represents the transition from one to the other. Averaging each pair of corresponding coordinates, for instance, gives an object halfway between start and finish. However, that's much too simple to be acceptable. It usually distorts the spaceship's shape.

The trick is to use rigid motions in space to do the interpolation. You might start by translating the ship to the midpoint, and rotating it through 45°. Do this again and it will be in the right finish location, having rotated by 90°. For the illusion of continuous movement, you can repeatedly translate by 1/90 of the difference in positions, and rotate by 1° each time. In practice much smaller steps would be used.

More abstractly, we can think of this procedure in terms of the 'configuration space' of all rigid motions. Each point in this space corresponds to a unique rigid motion, and nearby points give nearby motions. So a sequence of motions, each near to the previous one, corresponds to a sequence of points, each close to the previous one. Linking these points together in order we get a polygonal path in the space of rigid motions. Making the steps very small we get a continuous path. So now the problem of in-betweening from start image to finish image has been recast as that of finding a path through configuration space. If we want transitions to be smooth, this should be a smooth path, with no abrupt bends. There are good methods to smooth a polygon.

The 'dimension' of this configuration space – that is, the number of coordinates needed to define a point in it – is six. There are three dimensions' worth of translations: one coordinate for each of north–south, east–west, and up–down. Then we need two more to define the position of a rotation axis, and a final one for the angle of rotation. So what started out as a problem about moving an object smoothly in three dimensions has now become that of moving a point along a smooth path in six dimensions. This recast animation problem can be tackled using techniques from multidimensional geometry to design appropriate paths.

What's the Use?

*

In applied mathematics, the traditional way to handle rotations of a rigid object goes back to Euler. In 1752 he proved that any rigid motion that doesn't reflect the object is either a translation or a rotation about some axis.[54] However, for calculations he combined three rotations about the three axes in the usual coordinate representation of space, a method now called *Euler angles*. As an example, Shoemake considered the orientation of an aircraft, which in aeronautics is specified by three angles:

- Yaw (or heading) about a vertical axis, which gives the direction of the aircraft in a horizontal plane
- Pitch, rotation around a horizontal axis through the wings
- Roll, rotation around the line from nose to tail.

The first problem with this kind of representation is that the order in which the components are applied is vital. Rotations don't commute. The second is that the choice of the axes isn't unique, and different areas of application use different choices. A third is that the formulas for combining two successive rotations, expressed in Euler angles, is extremely complicated. These features don't cause too much trouble in basic aeronautical applications, which are largely about the forces acting on the aircraft when in a given orientation, but they're awkward for computer animations, where objects undergo entire sequences of movements.

Shoemake argued that quaternions, though less direct, provide a way to specify rotations that's far more convenient for animators, especially as regards in-betweening. A quaternion $a + b\mathrm{i} + c\mathrm{j} + d\mathrm{k}$ splits into a scalar part a and a vector part $v = b\mathrm{i} + c\mathrm{j} + d\mathrm{k}$. To rotate a vector v by a quaternion q, multiply v by q^{-1} on the left and by q on the right to get $q^{-1}vq$. Whatever q may be, the result is again a vector, with zero scalar part. Hamilton's rules for multiplying quaternions show, remarkably, that *any* rotation corresponds to a single quaternion. The scalar part is the cosine of half the angle through which the object rotates; the vector part points along the axis of rotation and has length equal to the sine of half that angle. So the quaternion neatly encodes the entire geometry of the rotation, with the slight inconvenience that the natural formulas work with half the angle, not the angle directly.[55]

Quaternions avoid distortions that can accumulate if an object is rotated many times, as it often must be. Computers can do exact calculations with whole numbers, but real numbers can't be represented with perfect accuracy, so tiny errors creep in. With the usual methods for representing transformations, the object being manipulated changes shape slightly, which the eye is good at spotting. In contrast, if you take a quaternion and change the numbers slightly, the result is still a quaternion, and it still represents a rotation, because *every* quaternion corresponds to some rotation. It's just a slightly different rotation from the exact one. The eye is less sensitive to such errors, and they can be compensated for easily if they get too big.

*

Quaternions are one way to create realistic motion in three dimensions, but so far what I've described applies to rigid objects. Spaceships, maybe; dragons, no. Dragons *flex*. So how do you make a realistic dragon in CGI? A common method applies not just to dragons, but to almost anything, and we'll work with a dinosaur because I've got suitable pictures. This approach reduces the motion of a flexible object to that of a set of linked rigid objects. You use whichever method you want for the rigid objects, with extra adjustments to link them together correctly. In particular, if you're using quaternions to rotate and translate rigid objects, the same methods can be adapted to work for a flexible dinosaur.

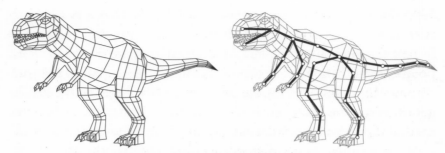

Left: Coarse polygon mesh *Tyrannosaurus rex. Right*:
Mesh attached to a rudimentary skeleton.

The first step is to create a three-dimensional digital model of the dinosaur, in which its surface is a complicated mesh of flat polygons – triangles, rectangles, less regular quadrilaterals. The software used for this task shows the shape geometrically, and you can move it, rotate, zoom in on details, and so on, with every movement shown on your computer screen. However, what the software manipulates is not the geometry as such, but a list of numerical coordinates for the points where the polygons meet. In fact, the mathematics that the software uses to help you draw the dinosaur is pretty much the same as that used to animate the result. The main difference is that at this stage the dinosaur is fixed and the viewpoint is being rotated and translated. In animation, the viewpoint can be fixed while the dinosaur moves; or, as with the swooping dragon, the viewpoint might move as well.

So now we've got ourselves a coarse, rigid dinosaur. How do we get him moving? What we don't do is what the artists had to do in the days of Mickey Mouse: redraw the image with the dinosaur in a slightly different position and repeat hundreds of times. We want the computer to do all the grunt work. So we reduce our dinosaur to a rudimentary skeleton: a small number of rigid rods ('bones') connected at their ends. We run these rods through the body, limbs, tail, and head. Not an anatomically correct skeleton, just a framework that lets us flex the main parts of the animal.

This skeleton is also represented as a list of coordinates for the two ends of each bone.

A very efficient way to get realistic movements, especially of people or humanoid creatures, is motion capture. An actor goes through the required motions in front of a camera, or several to get three-dimensional data. White dots are attached to key points on their body, such as feet, knees, hips, and elbows, and the computer analyses a video of the actor to extract how the dots move. The data from those are used to animate the skeleton. This is how Gollum was animated in the *Lord of the Rings* trilogy. Naturally, if you want weird inhuman movements (but realistic ones) then the actor has to move in a suitably weird way.

However we animate the skeleton, once we're happy with the result we 'drape' the mesh over the skeleton. That is, we combine the two lists of coordinates, specifying additional links between the positions of the bones and those of the surrounding bits of the mesh. Then, for a large part of the process, we forget about the mesh, and animate the skeleton. It's here that our work on rigid motions pays off handsomely, because each bone is rigid, and we want them to move in three dimensions. We must also impose constraints on the motion, so that the skeleton continues to hang together. If we move one bone, some ends of the bones attached to it must also move, so we translate those end coordinates to the right positions. Then we can move those bones rigidly too, which of course affects the bones attached to *them* ... and bone by bone we can make the whole skeleton flex slightly. We can move the feet to get him walking, flex his tail up or down or sideways, open his ferocious jaws – but we do it all on the skeleton. This is simpler, quicker, and *cheaper*, because the skeleton has many fewer pieces.

When we're happy that the skeleton is moving how we want, we might find it useful to drape the mesh back over it, starting with the first frame of the movement. The animation software then makes the mesh follow the skeleton's motions over successive frames, without us having to do any further work beyond a click or two of a mouse. By doing this we can check that the animation still looks realistic when the dinosaur follows what's happening to its skeleton.

Now we can play all sorts of creative games. We can move the 'camera' position, the viewpoint used by the software, zooming in for a close-up, viewing the running dinosaur from a distance, whatever. We can create other creatures, maybe a herd of herbivores fleeing from the monstrous tyrannosaur. Again this is done initially with skeletons and then meshes get draped over them. We can animate each creature separately, then fit them all together to create a hunting scene.

Since skeletons are just stick figures, at this point we may not have done anything to prevent two creatures occupying the same space. More software tweaks can alert us to any collisions of this kind. When we drape the meshes over the skeletons, polygons that are in the front will overlap those behind, and since dinosaurs aren't transparent, we have to eliminate any regions that ought to be hidden. All this is done using simple calculations in coordinate geometry, but rather a lot of calculations. Until computers got really fast, it wasn't feasible. Now it's routine.

There's still more work to do, because a dinosaur that looks like a lot of polygons isn't terribly impressive. We have to drape realistic skin patterns over the polygons, then sort out colour information, maybe create realistic textures – fur looks very different from scales. Each step requires different software, implementing different mathematical techniques. This step is called rendering, and it assembles the final picture that appears on the screen when we watch the movie. But at the heart of everything are billions of calculations that move points and edges around rigidly.

These mathematical methods have another advantage, too. At any stage we can decide something's not quite right, and change it. If we want a green dinosaur instead of a brown one, we don't have to draw everything all over again. We use the same skeleton and mesh, the same movements, and the same skin texture, but change the colour.

When animating a movie or a game, teams of experts use a range of standard software packages that the industry has developed to carry out these processes. To give you some idea of the complexity of these activities, I'll take a look at some of the companies and software packages used to make the movie *Avatar*.

The bulk of the animation was done by Weta Digital in New Zealand, famous for its work on *Lord of the Rings* and *The Hobbit*. Industrial Light & Magic, founded in 1975 by George Lucas to create special effects for the first *Star Wars* movie, created 180 such sequences, mainly the aircraft in the final battle. The rest, in the UK, Canada, and the USA, added vital special details such as control-room screens and 'heads-up' displays on visors, simulating future technology. Autodesk Maya took care of most such shots. Luxology Modo was used for model design, in particular the *Scorpion*. Houdini created the Hell's Gate scenes and interiors. The alien creatures were designed using ZBrush. Autodesk Smoke performed colour correction, Massive simulated alien vegetation, and Mudbox took care of the floating mountains. Initial concept art and textures were created using Adobe Photoshop. In all, about a dozen companies were involved, using 22 different software tools plus innumerable specially coded plug-ins.

*

Some very sophisticated mathematics is now being stirred into the mix for CGI animation. The aims, always, are to make the animator's task as simple as possible, to obtain realistic results, and to keep costs and time down. We want it all, we want it now, and we want it cheap.

Suppose, for instance, that the movie studio has a library of animations of a dinosaur going through various movement sequences. In one, it gallops forwards through one 'gait cycle', one segment of a periodically repeating motion. In another, it leaps into the air and crashes down. You want to create a sequence in which it gallops after a small herbivore, and leaps on top of it. An efficient way to get started would be to string together a dozen or so gait cycles of the gallop, and then add the leap at the end. Of course you would then tweak everything so it's not obvious that the same animation is being repeated a dozen times, but this is a good start.

It makes sense to string the sequences together at the skeleton level. All of the other stuff like draping meshes and adding colour and texture can be done later. So you do the obvious, and join twelve copies of the gallop cycle to a leap, and see what it looks like.

It looks awful.

The separate bits are OK, but they don't fit together smoothly. The result is jerky and unconvincing.

Until recently, your only recourse would have been to modify the joins by hand, maybe interpolating some new bits of movement. Even then, it would have been a tricky business. Some recent developments in the mathematical techniques promise to solve the problem in a far better way. The idea is to use smoothing methods to fill in any gaps and iron out abrupt transitions. The key step is to find good ways to do that with a single bone in the skeleton, or more generally with a single curve. Having solved that problem, you can stitch the skeleton together from the individual bones.

The area of mathematics that's currently being tried is called shape theory. So let's start with the obvious question: what is a shape?

In ordinary geometry we encounter a lot of standard shapes: triangle, square, parallelogram, circle. When these shapes are interpreted in coordinate geometry, they turn into equations. In the plane, for instance, the points (x, y) on a unit circle are precisely those that satisfy the equation $x^2 + y^2 = 1$. Another very convenient way to represent the circle is to use a so-called *parameter*. This is an auxiliary variable, say t, which we can think of as time, together with formulas for how x and y depend on t. If t runs through some range of numbers, each value of t gives two coordinates $x(t)$ and $y(t)$. Get the formulas right, and those points define the circle.

The standard parametric formulas for a circle are trigonometric:

$$x(t) = \cos t, \quad y(t) = \sin t$$

However, it's also possible to change how the parameter appears in the formula and still get a circle. For instance, if we change t to t^3, then

$$x(t) = \cos t^3, \quad y(t) = \sin t^3$$

also determines a circle – and it's the same circle. This effect happens because the time parameter conveys more information than just how x and y vary. For the first formula, this point moves with a constant speed as t varies. For the second, it doesn't.

Shape theory is a way to get round this lack of uniqueness. A shape is a curve, considered as an object that doesn't depend on a particular parametric formula. So two parametric curves define the same shape if you can change the parameter and convert one formula to the other, like changing t to t^3. Over the past century, mathematicians have come up with a standard way to do that sort of thing. It's not what anyone else is likely to have thought of, because it requires a rather abstract point of view.

The first step is to consider not just one parametric curve, but the 'space' of *all possible* parametric curves. Then we say that two 'points' in this space (that is, two parametric curves) are equivalent if you can get from one to the other by changing the parameter. A 'shape' is then defined to be an entire equivalence class of curves – the set of all curves equivalent to a given one.

It's a more general version of the trick used to do arithmetic to a modulus. In the integers modulo 5, for instance, the 'space' is all integers, and two integers are equivalent if their difference is a multiple of 5. There are five equivalence classes:

> All multiples of 5
> All multiples of 5 with 1 added
> All multiples of 5 with 2 added
> All multiples of 5 with 3 added
> All multiples of 5 with 4 added.

Why stop there? Because a multiple of 5 with 5 added is just a slightly larger multiple of 5.

In this case the set of equivalence classes, which is denoted $\mathbb{Z}_5$, has a lot of useful structure. Indeed, Chapter 5 showed that much basic number theory hinges upon just this structure. We say that $\mathbb{Z}_5$ is the 'quotient space' of the integers modulo 5. It's what you get if you pretend that numbers differing by 5 are identical.

Something similar happens to get shape space. Instead of the integers, we have the space of all parametric curves. Instead of changing numbers by multiples of 5, we change the parameter formula. So we end up with a 'quotient space' that's the space of all parametric curves modulo changes in the parameter. That probably sounds meaningless, but it's a standard trick whose value has become apparent over a long period of time. One reason it's valuable is that the quotient space is the natural description of the objects we're interested in. Another is that usually the quotient space inherits interesting structure from the original one.

For the space of shapes, the main interesting item of structure is a measure of the distance between two shapes. Take a circle and deform it slightly: you get a closed curve, which stays close to the circle, but it's different. Deform the circle a lot, and you get a closed curve that intuitively is *more* different: 'further away'. This intuition can be made precise, and it can be proved that shape space has a sensible and natural concept of distance: a metric.

Once a space has a metric you can do all kinds of useful things. You can, in particular, distinguish continuous changes from discontinuous ones, and you can up the ante a bit to distinguish smooth changes from non-smooth ones. And finally we're back with the problem of stitching animation sequences together. At the very least, this metric on shape space lets us detect discontinuities or lack of smoothness *on the computer*, by doing the sums, rather than by eye. But there's more.

Mathematics has many smoothing techniques, which can transform a discontinuous function into a continuous one, or a non-smooth one into a smooth one. It's been discovered that you can apply those techniques to shape space. So a stitched-together sequence with a sudden discontinuity can be modified automatically to get rid of that discontinuity, by the computer doing the right sums. This isn't easy, but it can be done, and it can be done efficiently enough to save money. Merely calculating the distance between two curves uses optimisation methods, a bit like the ones we encountered in the Travelling Salesperson Problem. Smoothing out a sequence involves solving a differential equation rather like Fourier's equation for the flow of heat, which we meet in Chapters 9 and 10. Now an entire animated sequence of curves is persuaded to 'flow' into a different animated sequence, smoothing out discontinuities – which again is like heat flow smoothing out a square wave.[56]

Similar abstract formulations also make it possible to convert animations into similar but different ones. A sequence showing a dinosaur walking can be tweaked to make the animal run. It's not just a matter of speeding up the action, because the way an animal runs is visibly different from how it walks. This methodology is still in its infancy, but it strongly suggests that some very high-level mathematical thinking could pay off big time in future movie animations.

These are just some of the ways that mathematics contributes to animation. Others create simplified versions of physical processes to simulate waves on the ocean, snowdrifts, clouds, and mountains. The aim is to get realistic results while keeping the calculations as simple as possible. There are now extensive mathematical theories about representing human faces. In *Rogue One*, part of the *Star Wars* series, actors Peter Cushing (who had died in 1994) and Carrie Fisher (who had died in 2016) were recreated digitally by draping their faces over those of body doubles. It wasn't terribly convincing, and fans objected vociferously. *The Last Jedi* used a better method: choose out-takes of Fisher from previous movies and string them together, adapting the script to fit. However, a lot of CGI was still needed to change her clothing, for consistency. In fact, almost everything except her face was rendered digitally – head, hairstyle, body, clothes.[57]

The same techniques are already being used to create Deep Fakes as political propaganda. Film someone making racist or sexist remarks, or appearing to be drunk; then drape your opponent's face over the top and put it on social media. Even when the fakery is detected, you're well ahead of the game, because rumours travel faster than facts. Mathematics, and the technology that depends on it, can be used for bad as well as for good. What matters is how we use it.

8

Boing!

A spring is an elastic object that recovers its original shape when released after being pressed or pulled. It is used to store mechanical energy by exerting constant tension or absorbing movement. Springs are used in virtually every industry, ranging from the automotive industry and construction to furniture.

Confederation of British Industry, *Product Factsheet: Springs in Europe*

Recently we bought a new mattress. The one we chose has 5,900 springs. The cross-sectional diagram in the department store showed densely packed arrays of loosely coiled springs, with a layer of smaller springs on top. The really high-end mattresses had another 2,000 springs, *inside* the main layer. Today's technology is a far cry from the days when a mattress would have about 200 rather large and not very comfortable springs.

A spring is one of those gadgets that are ubiquitous but seldom noticed – until they go wrong. There are valve springs in car engines, long thin springs in retractable ball-point pens, and springs of many different shapes and sizes in computer keyboards, toasters, door handles, clocks, trampolines, sofas, and blu-ray players. We don't notice them because they're tucked away inside our appliances and furniture, out of sight and consequently out of mind. Springs are big business.

Do you know how springs are made? I certainly didn't until 1992, when my office phone rang.

'Hello? This is Len Reynolds. I'm an engineer at the Spring Research and Manufacturers Association in Sheffield. I've been reading your book on chaos theory, and you mention a method to find the shape of a chaotic attractor from observations. I think it might help to solve a problem we've had in the spring-making industry for the last twenty-five years. I tried it out on some test data, with my ZX81.'

The Sinclair ZX81 was one of the first mass-market home computers, using a TV set as a display and a cassette tape to store software. It was about the size of a book, made of plastic, with a magnificent 1K of memory. You could plug an extra 16K in the back, provided you took precautions to stop it falling out. I made a wooden frame to hold the RAM pack in place; other people used Blu Tack.

It wasn't exactly state-of-the-art computational technology, but Len's preliminary results were sufficiently promising to secure a grant of £90,000 (then about $150,000) from the Department of Trade and Industry (DTI), with matching funds (in kind, not cash) from a consortium of spring and wire manufacturers. The money paid for a three-year project to improve quality control testing of spring wire, which led to two other projects over a five-year period. At one stage it was estimated that the outcome could save the spring and wire industries £18 million ($30 million) a year.

There are literally thousands of such applications of mathematics to industrial problems, going on all the time, mostly under the radar. Many are commercial secrets, protected by non-disclosure agreements. From time to time UK organisations such as the Engineering and Physical Sciences Research Council or the Institute of Mathematics and its Applications publish brief case studies of a few of these projects, and the same happens in the USA and elsewhere. Without these projects, and many other targeted uses of mathematics by corporations large and small, worldwide, none of the appliances and devices we use every day would exist. Yet it's a hidden world, and few of us even suspect it's there.

In this chapter I'll take the lid off the three projects that I participated in. Not because they're especially important, but because I know what they involved. The essential ideas were published, mostly in industry journals, and are public domain. My aim is to show you that the way mathematics gets used in industry is often indirect and surprising, with a dash of serendipity.

Like Len's phone call.

*

The problem that had puzzled the wire and spring industries for a quarter of a century was simple and basic. Springs (made by spring-making companies) take wire (made by wire manufacturers) and run it through coiling machines to make springs. Most wire performs perfectly well, producing springs in the correct ranges for size and springiness. But from time to time, a consignment of wire refuses to coil correctly, even in the hands of a highly skilled machinist. The usual quality control methods of the early 1990s couldn't tell good wire from bad. Both passed the same tests for chemical composition, tensile strength, and so on. Visually, bad wire looked just like good wire. But when good wire was fed into a coiling machine, what came out was the spring you wanted; when bad wire was fed in, what came out either looked like a spring but had the wrong size, or in the worst cases was just a hopeless tangle.

Trying to coil the wire wasn't an efficient or effective test. If the wire was bad, it would tie up an expensive coiling machine for a couple of days before the operator was convinced that that batch of wire would never make springs. Unfortunately, since the wire had passed the usual tests, the manufacturer could reasonably maintain that there was nothing wrong with it: something must have been wrong with the set-up on the coiling machine. Both industries deplored the resulting Mexican stand-off, both wanted a reliable way to find out who was right, and both were willing to discover that it wasn't them. The goodwill was there, but they needed an objective test.

When we started the project, one of the first steps was to take the mathematicians to a spring-making company and show them how wire becomes springs. It's all about geometry.

The commonest springs are compression springs. Push their ends closer together, and they push back. The simplest design is a helix, like a spiral staircase. Imagine a point going round and round a circle at uniform speed; now displace it at right angles to the circle at uniform speed. The curve it traces out in space is a helix. For practical reasons helical springs often close up at both ends, as if the moving point first goes round a circle in the plane before starting to move at right angles to it, and then stops moving in that direction for the final coil. This protects the coil from getting its ends caught in things, and also protects people against being the thing that the ends get caught in.

Mathematically, a helix is characterised by two properties, its curvature and its torsion. Curvature measures how sharply or gently it bends. Torsion measures how much it twists out of the plane determined by the direction in which it's bending. (Obviously there's a technical definition, but let's not get tangled up in the differential geometry of space curves.) For a helix, both of these quantities are constant. So when you look at the helix from the side, the coils are evenly spaced and tilted at the same angle – that comes from the constant speed along the axis of the helix. When you look from one end, all the coils align with each other to create a circle: that's the uniform motion round and round. A small circle corresponds to high curvature, a big one to low curvature; a steeply rising helix corresponds to high torsion, a slowly rising one to low torsion.

A coiling machine embodies these properties mechanically,

in a wonderfully simple manner. The coiling machine feeds wire from a big, loose reel called a swift, past a small tool, just a rigid piece of metal. This simultaneously bends the wire in one direction and gives it a small push at right angles to that. The bending creates curvature and the push creates torsion. As the wire keeps feeding through, the machine spins off turn after turn of the helix. When it's long enough, another tool cuts it off, ready to form the next spring. Extra gadgetry changes the torsion to zero near each end to flatten those coils and close them up. The process is fast – several springs per second. One manufacturer made tiny springs from special wire at a rate of 18 per second on each coiling machine.

Wire and spring companies are generally fairly small, technically Small and Medium-sized Enterprises (SMEs). They're sandwiched between very large suppliers, such as British Steel, and very large customers, such as automotive and mattress companies, so their profit margins are squeezed at both ends. To survive, they have to stay efficient. No individual company can justify the cost of having its own research department, so SRAMA (the Spring Research and Manufacturers Association, since renamed the Institute of Spring Technology, IST) is a kind of joint research and development operation, a collaborative venture funded by its member companies. Len and his colleagues at SRAMA had already made some progress on the coiling problem, based on what goes wrong. The curvature and torsion of the growing coil depend on material properties of the wire, such as its plasticity: how easy or hard it is to bend it. When the coil forms a nice regular helix, these properties are uniform along the wire. When it doesn't, they're not. So it seemed likely that poor coilability stemmed from irregular variation of these material properties along the wire. The question then became: how to detect such variations.

The answer was to force the wire to make a coil by wrapping it round a metal bar, much like spaghetti being wrapped round a fork. Then you can measure the spacings between successive coils. If they're all fairly equal, it's good wire. If they're all over the place, it's bad wire. Except that sometimes they could vary quite a lot and the wire would still make springs. Maybe not as accurately as really good wire, but good enough for some applications. So the heart of the problem was: how do you quantify – put a number to – the extent to which the wire is 'all over the place'?

SRAMA's engineers applied all the usual statistical tools to the list of measurements, but nothing corresponded terribly closely to coilability. That's where my book on chaos theory came in.

*

Chaos theory, a name invented by the media, is better known to mathematicians as part of the broader theory of nonlinear dynamics, which is about how systems behave when their behaviour over time is governed by a specific mathematical rule. Measure the state of the system *now*, apply the rule, and deduce the state a tiny period of time into the future. Then repeat. As time ticks on, you can calculate the state of the system as far as you like into the future. This technique is dynamics. Roughly, 'nonlinear' means that the rule doesn't just make the future state proportional to the current one, or to the difference between the current one and some reference state. For continuously varying time, the rule is specified by a differential equation, which relates the rate of change of the system's variables to their current values.

There's also a discrete version in which time ticks step by step, described by a difference equation: state after one tick is what happens to the current state when you apply the rule. It's the discrete version that solves the coiling problem. Fortunately, that's the easier one to understand. It works like this:

$$\text{state at time } 0 \rightarrow \text{state at time } 1 \rightarrow \text{state at time } 2 \rightarrow \dots$$

where the arrow means 'apply the rule'. For instance, if the rule is 'double the number' and we start with the initial state equal to 1, then successive steps produce the sequence of states 1, 2, 4, 8, ..., doubling each time. This is a linear rule because the output is proportional to the input. A rule like 'square and subtract 3' is nonlinear, and in this case it produces the sequence of states

$$1 \to -2 \to 1 \to -2 \to \ldots$$

which repeats the same two numbers over and over again. This is 'periodic' dynamics, much like the cycle of the seasons, say. The future behaviour is completely predictable, given the initial state: it just alternates between 1 and -2.

On the other hand, if the rule is 'square and subtract 4', we get

$$1 \to -3 \to 5 \to 21 \to 437 \to \ldots$$

and the numbers just get bigger and bigger (except for the second one). The sequence is still predictable: just keep applying the rule. Since it's deterministic – it has no random features – each successive value is uniquely determined by the previous one, so the *entire future* is completely predictable.

The same goes for continuous-time versions, although the predictability is not as obvious in that case. A sequence of numbers of this kind is called a time series.

Inspired by Galileo Galilei and Newton, mathematicians and scientists uncovered innumerable rules of this type, such as Galileo's rule for the position of a body falling under gravity and Newton's law of gravitation. This process led to the belief that any mechanical system obeys deterministic rules, so it's predictable. However, the great French mathematician Henri Poincaré discovered a loophole in this argument, which he published in 1890. Newton's law of gravity implies that two celestial bodies, such as a star and a planet, move in elliptical orbits about their common centre of mass, which in this case is usually inside the star. The motion is periodic, the period being the time it takes to orbit once and come back to the starting position. Poincaré investigated what happens if there are three bodies (sun, planet, moon), and found that in some cases the motion is extremely irregular. Later mathematicians, belatedly following up on this discovery, realised that this type of irregularity makes the future of such a system unpredictable. The loophole in the 'proof' of predictability is that it's valid only when you can measure the initial state and do all of the calculations with perfect accuracy – correct to infinitely many decimal places. Otherwise, even very tiny discrepancies can grow exponentially fast, until they swamp the true value.

This is chaos, or more properly, deterministic chaos. Even when you know the rules, and they have no random features, the future might not be predictable in practice, even if it's predictable in theory. In fact, the behaviour can be so irregular that it looks random. In a truly random system, the current state provides no information about the next state. In a chaotic system, there are subtle patterns. The secret patterns behind chaos are geometric, and can be visualised by plotting solutions of the model equations as curves in the space whose coordinates are the state variables. Sometimes, if you wait a while, those curves start to trace out a complex geometric shape. If curves from different starting points all trace out the same shape, we call the shape an attractor. The attractor characterises the hidden patterns in the chaotic behaviour.

A standard example is the Lorenz equations, a continuous-time dynamical system that models a convecting gas, such as hot air in the atmosphere. This equation has three variables. In a plot of how they change, using a three-dimensional coordinate system, the solution curves all end up travelling along a shape rather like a mask: the Lorenz attractor. Chaos arises because although the solution curves wander around on (well, very close to) this attractor, different solutions wander around in very different ways. One might (say) wrap six times round the left-hand loop and then

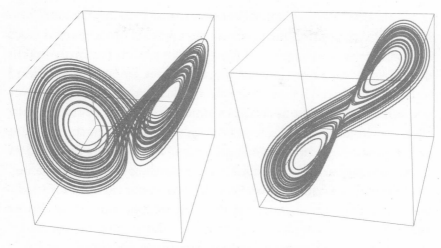

Left: The Lorenz attractor. *Right*: A reconstruction
of its topology from a single variable.

seven times round the right-hand one; a nearby curve might wrap
eight times round the left-hand loop and then three times round
the right-hand one, and so on. So the predicted futures of these
curves are very different, even though they start from very similar
values of the variables.

However, short-term predictions are more reliable. At first,
two neighbouring curves stay close together. Only later do they
start to diverge. So a chaotic system is predictable in the short run,
unlike a truly random system, which isn't predictable at all. This
is one of the hidden patterns that distinguish deterministic chaos
from randomness.

When we work with a specific mathematical model, we know
all the variables and can use a computer to calculate how they
change. We can visualise the attractor by plotting these changes
in the coordinates. When observing a real system that might be
chaotic, that luxury isn't always available. In the worst case, we
may be able to measure only one of the variables. Since we don't
know the others, we can't plot the attractor.

This is where Len's insight comes into play. Mathematicians have devised cunning methods to 'reconstruct' an attractor from measurements of a single variable. The simplest is Packard–Takens or sliding-window reconstruction, developed by Norman Packard and Floris Takens. It introduces new 'fake' variables by measuring the same variable but at different times. So instead of the original three variables at simultaneous times, we look at just one variable within a window three time-steps long. Then we slide the window along one step and do the same, repeating the process many times. The right-hand figure shows how this works for the Lorenz attractor. It's not the *same* as the left-hand figure, but unless you choose the time-step very badly, the two pictures have the same topology: the reconstructed attractor is a continuously distorted version of the actual one. Here, both images look like masks, with two eyeholes, but one is a twisted version of the other.

This technique provides a qualitative picture of the attractor, which tells us what *kind* of chaos to expect. So Len, wondering whether the same trick would work on his spring data, made a two-dimensional plot treating successive gaps between coils as a time series and applying sliding-window reconstruction. He didn't get a crisp geometric shape like a mask, though; what he got was a fuzzy cloud of points. This suggested that the sequence of gaps might not have been chaotic in the technical sense used by mathematicians.

So the method was no use?

Not at all.

What captured Len's attention was the overall *shape* of that fuzzy cloud. The wire samples had been tested laboriously on a coiling machine, so he knew which samples were good, bad, or indifferent. Could the reconstructed point cloud tell which was which? Apparently, it could. When the wire was really good, coiling easily and making very precise springs, the cloud was small and roughly circular. When the wire was acceptable, coiling fairly easily but making springs with more variable sizes, the cloud was larger, but still roughly circular. In contrast, when the wire was bad, impossible to coil into springs, the cloud was long and thin, like a cigar.

If the same pattern persisted for other samples, you could skip the slow, expensive test on a coiling machine, and use the shape and size of the fuzzy cloud to characterise good, indifferent, and bad wire. That would solve the practical problem of finding a cheap, effective test for coilability. It doesn't actually *matter* whether the spacing of the coils is random, chaotic, or a bit of both. You don't have to know exactly how the material properties vary along the wire, or even what those properties are. You certainly don't have to do very complicated calculations in elasticity theory, verified by equally complicated experiments, to understand how those variations translate into good or bad coilability. All you need to know is how the sliding-window plot distinguishes good wire from bad, and you can check that by testing it on more wire samples and comparing with how it performs on a coiling machine.

The reason why standard statistical measures on the data, such as the mean (average) and variance (spread) were no help now became clear. Those measures ignore the order in which the data arise: how each spacing relates to the previous one. If you shuffle the numbers, the mean and variance don't change, but the shape of the point cloud can change dramatically. And that, very likely, is the key to making good springs.

To investigate this insight we built a quality control machine, FRACMAT, which wound a test coil round a metal rod, scanned it with a laser micrometer to measure the successive gaps, fed those numbers into a computer, applied sliding-window reconstruction to get a cloud of points, estimated the best-fitting ellipse to see whether it was circular or cigar-shaped and how big it was, and worked out how good or bad the wire sample was. It was a practical application of chaos *theory*, the reconstruction method, to a problem that probably wasn't even chaotic in a technical sense. Fittingly, the DTI funding wasn't for research, but for technology transfer; we transferred the reconstruction method from the mathematics of chaotic dynamics to time series of observations of a possibly non-chaotic real-world system. Which is exactly what we told them we were going to do.

*

Chaos isn't just a fancy word for 'random'. Chaos is predictable in the short term. If you roll a dice (yes, I know it's 'die', but most of us say 'a dice' nowadays and I've stopped fighting that) then the current throw tells you nothing about what will happen next. Whatever I throw this time, all of the numbers 1, 2, 3, 4, 5, 6 are equally likely next time. Assuming the dice is fair, rather than loaded to make some numbers more likely. Chaos is different. If chaos were a dice, there would be patterns. Maybe a throw of 1 could be followed only by 2 or 5, while a throw of 2 could be followed only by 4 or 6, and so on. The next result can be predicted to some extent, but the fifth or sixth throw from now could be anything. The further ahead you want to know about, the more uncertain the prediction gets.

The second project, DYNACON, grew out of the first when we realised that it might be possible to exploit this short-term predictability of chaos to control a coiling machine. If we could somehow measure the lengths of springs as they were produced, and spot trends in the figures that suggested the machine really was acting chaotically, it might be possible to see the bad springs coming and adjust the machine to compensate. Manufacturers had already found ways to measure the length as the spring was made, to divert inaccurate springs into a separate bin, but we wanted more. Not just to sort out the bad ones as they were made, but to stop bad ones being made at all. Not perfectly, but enough to avoid wasting lots of wire.

Most mathematics deals in precision. A number is (or is not) equal to 2. That number does (or does not) belong to the set of primes. The real world is often much fuzzier. A measurement may be close to 2 but not exactly equal to it; moreover, if you measure the same quantity again the result may be slightly different. Although a number can't be 'almost prime', it can certainly be 'almost an integer'. That description is reasonable for a number like 1·99 or 2·01, say. In 1965 Lotfi Zadeh and Dieter Klaua independently formulated a precise mathematical description of this kind of fuzziness, known as fuzzy set theory, together with a related concept of fuzzy logic.

In conventional set theory, an object (such as a number) either belongs to a specified set or it doesn't. In fuzzy set theory, there's a precise numerical measure of the *extent* to which it belongs. So the number 2 might half-belong to the set, or one-third-belong to it. If that measure is 1, the number definitely belongs to the set, and if it's 0 the number definitely doesn't. With only 0 and 1 we have conventional set theory. If we allow any measure between 0 and 1, the degree of fuzzy membership captures the grey area between those extremes.

Some prominent mathematicians rushed to dismiss the idea, either claiming that fuzzy set theory is just probability theory in disguise, or arguing that most people's logic is fuzzy enough without insisting that mathematics should go the same way. Quite what motivates some academics to be so instantly dismissive about new ideas baffles me, especially when their reasons for doing so don't make any sense. No one was suggesting *replacing* standard logic by fuzzy logic. It was just being offered as another weapon in the armoury. Although fuzzy sets superficially look like probability, the rules are different, and so is the interpretation. If a number belongs to a set with probability 1/2, and you're a frequentist, you're saying that if we repeat the experiment many times, the number will be in the set about half the time. If you're a Bayesian, your confidence that the number belongs to the set is 50%. But in fuzzy set theory, there's no chance element. The number is definitely *in* the set – but the extent to which it belongs isn't

1. It's *precisely* 1/2. As for the gibe about poor logic: fuzzy logic has precise rules, and any argument that uses it is either correct or not – depending on whether it has obeyed those rules. I guess the word 'fuzzy' led some people to assume, without bothering to find out, that the rules themselves were malleable and poorly defined. Not so.

A different issue, which no doubt muddied the waters, is the extent to which fuzzy sets and fuzzy logic add anything of value to mathematics. It's all too easy to set up extensive formal systems that are little better than pretentious jumbles of content-free formulas – 'abstract nonsense'. It was, I suspect, all too tempting to view Zadeh's brainchild in that light, especially since the basics were hardly deep or difficult. Now, the proof of the pudding may be in the eating alone, but the value of mathematics can be assessed in several ways, only one of which is its intellectual depth. Another, rather relevant to this book, is *utility*. And many almost trivial mathematical ideas have turned out to be extraordinarily useful. Decimal notation, for instance. Brilliant, innovative, clever, game-changing – but not deep. A child can understand it.

Fuzzy logic and fuzzy set theory probably fail the depth criterion, at least by comparison with the Riemann Hypothesis or Fermat's Last Theorem. But they've proved very useful indeed. They come into their own whenever we're not entirely sure about the accuracy of the information we're observing. Fuzzy mathematics is now widely used in areas as diverse as linguistics, decision making, data analysis, and bioinformatics. It's used when it does the job better than any alternative, and we can safely ignore it when it doesn't.

I don't want to go into the details of fuzzy set theory, which aren't really necessary to appreciate our second project. We tried several methods to predict when the coiling machine was about to produce bad springs, adjusting the machine accordingly. One method is known in the trade as a Takagi–Sugeno fuzzy identifier model, after engineers Tomahiro Takagi and Michio Sugeno.[58]

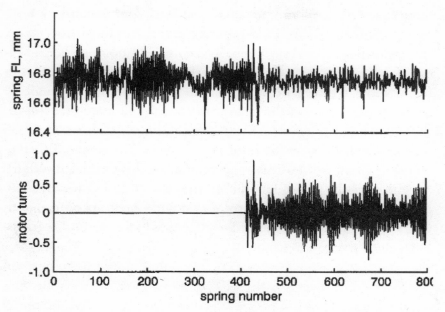

Effect of switching on the fuzzy self-tuning controller. The spring
number runs from left to right. *Top*: the measured lengths of the
springs. *Bottom*: Activity of the controller, measured by the number
of times the controlling motor turns. Springs 1–400 are without
control, and the variability in the length is high. Springs 401–800 were
coiled with the controller on: the variability is visibly smaller.

This implements, in the precise formalism of fuzzy mathematics, systems of rules that are themselves fuzzy. In this case, the
rules take the form 'if the (necessarily fuzzy) measurement of
the current spring's length is X, then do Y to adjust the coiling
machine'. The rules also take into account the previous adjustment, together with an estimate of disturbances caused by variable
material properties of the wire, wear of the machine tool, and so
on. All the data are fuzzy, and so are the actions you take; the
mathematical formalism handles this automatically to adjust the
coiling machine on the fly.

For our strip metal project, we tried three different control methods. First, we ran the machine with the control system switched off, to establish a baseline from which to judge how effective any other control system is. The data obtained also helped to estimate various parameters in mathematical models. Next, we ran it with an integral controller, which uses a fixed mathematical formula to predict the change in adjustment from one coil to the next. Finally, we used fuzzy self-tuning control, which fine-tunes its own rules on the hoof according to the observed spring lengths. When we did this with carbon steel wire, the standard deviation of the spring lengths – a measure of how variable they are – was 0·077 with no control, 0·065 with integral control, and 0·039 with fuzzy self-tuning. So the fuzzy logic method worked best, and halved the variability.

*

Another basic principle in mathematics is that once you've found something that does the job, you work it to death. An idea of proven value can often be exploited in related, but different, circumstances. Our third project, also part of DYNACON, went back to FRACMAT, but we modified the test device to suit an industry that's similar to springmaking but uses strip metal rather than wire.

You'll almost certainly have some products made by the strip industry in your home. In the UK, every electric plug contains a fuse, held by copper clips. Those clips are made from a large coil of thin, narrow, copper strip. A machine feeds strip metal past a series of tools arranged in a rough circle, all pointing towards the centre where the strip runs through. Each tool makes one bend in the strip, at a specific angle and position, punches a hole, or carries out any other operation that's required. Finally a cutting tool chops off the finished clip, which drops into a bin. A typical machine can make ten or more clips every second.

The same process is used to make a huge range of small metal objects. One UK company specialises in making the clips that hold the supports for suspended ceilings together, churning out hundreds of thousands every day. Just as the spring makers have problems assessing whether wire will coil well, so the clip makers have problems assessing whether a given sample of strip will bend the way it's intended to. The source of the problem is similar: variable material properties, such as plasticity, along the length of the strip. So it seemed reasonable to try the same method of sliding-window reconstruction on strip metal.

However, it's not sensible to try to force strip metal to make a coil. It's the wrong shape to do this easily, and coiling has little connection to the way clips are made. The key quantity is how much the strip bends for a given applied force. So, after much thought, we redesigned the testing machine, and came up with something much simpler. Feed the strip between three rollers, so that the one in the middle forces it to bend. Let the roller in the middle move a little, on a hard spring, and measure how far it moves as the strip passes beneath it. The strip gets bent and then flattened out again, and you can measure the force required to bend it. If the plasticity of the strip varies along its length, so does this force.

Instead of discrete measurements of coil spacings for wire, made by a laser micrometer, we now have continuous measurements of forces. The machine also measures surface friction, which turned out to have an important effect on quality. The data analysis is much the same, though. This test machine is smaller than FRACMAT, simpler to build, and as a bonus the test is non-destructive: the strip is returned to its initial state, and could be used in manufacturing if you wanted.

*

What did we learn?

We probably saved the wire and spring industries quite a lot of cash, so we learned that this kind of mathematical data analysis has dollars-and-cents value. To some extent, the mere existence of FRACMAT persuaded the wire makers to improve their production procedures, which in turn helped the spring makers. The test machines are still in use, and IST continues to act as a common resource for numerous small companies, doing the testing for them.

We learned that sliding-window reconstruction can be useful even when the data are not known to be generated by some nice, clean, mathematically precise, chaotic dynamic. Do material properties of wire vary chaotically in the technical sense? We don't know. We didn't *need* to know to create the new test procedure and machine. Mathematical methods aren't confined to the particular context for which they were originally developed. They're portable.

We learned that sometimes when you try to transfer a trick that works into a new context – control – it doesn't do the job. Then you have to look for different methods that do – fuzzy logic.

We learned that sometimes this kind of transfer works really well. Better, in some ways, than the first attempt. Our machine for strip metal also works on wire, and is non-destructive.

Above all, we learned that when a team of people with very different expertise join forces on a common problem, they can solve it in ways that no single team member can do on their own. As humanity rolls on into the twenty-first century, facing new and mutually interacting problems on every level, from the social to the technological, that's a very important lesson.

9

Trust Me, I'm a Transform

A patient visited a doctor for the first time.

'Who did you consult before coming to me?' the doctor asked.

'The village pharmacist.'

'And what foolish advice did that numbskull give you?'

'He told me to come and see you.'

Author unknown

The manner in which the author arrives at these equations is not exempt of difficulties and that his analysis to integrate them still leaves something to be desired on the score of generality and even rigour.

Report on Joseph Fourier's submission for the 1811 Mathematics Prize of the Paris Institute

Nowadays, a visit to a hospital often involves a scan. There are many kinds of scanner: magnetic resonance imaging, PET scans, ultrasound ... Some show moving images in real time, some use computer trickery (that is, mathematics) to provide three-dimensional images. The most remarkable feature of these technological marvels is that they show pictures of what's going on *inside* your body. Not so long ago, that would have been considered magic. It still looks like it.

In ancient times, which in this case means anything before

1895, doctors had to use their own senses to investigate what was ailing their patients. They could feel the body to get some idea of the shape, size, and position of some of the internal organs; they could listen to the heartbeat and feel the pulse; they could assess temperature, and smell, feel, and taste bodily fluids. But the only way they could find out what the inside of the human body really looks like was to cut it open. Sometimes they couldn't even do that, because religious authorities often forbade dissection, even though it was quite common on the battlefield, though not with any medical objective in mind. Those same authorities often approved of that kind of dissection, provided it was used against people with different beliefs.

A new era began on 22 December 1895, when the German physics professor Wilhelm Röntgen photographed his wife's hand, obtaining an image that showed the bones of her fingers. It was black-and-white, like virtually all photographs at the time, and rather blurred, but the ability to look inside a living body was unprecedented. She was unimpressed. On seeing the photo of part of her skeleton, she said: 'I have seen my death.'

Röntgen's discovery was pure serendipity. In 1785 an actuary called William Morgan carried out some experiments in which he passed an electric current through a partial vacuum in a glass tube. This produced a faint glow, best seen in the dark, and he presented his results to the Royal Society of London. By 1869 physicists experimenting in the now-fashionable field of discharge tubes had noticed a strange new type of radiation, called cathode rays because they were emitted by the tube's cathode (negative electrode). In 1893 Fernando Sanford, a physics professor, published an article on 'electric photography'. He constructed a tube with a thin aluminium sheet at one end, and cut a hole in it. When the current was switched on, whatever caused the faint glow passed through the hole, hit a photographic plate, and reproduced the shape of the hole. His discovery was reported in the press – the *San Francisco Examiner*'s headline was: 'Without

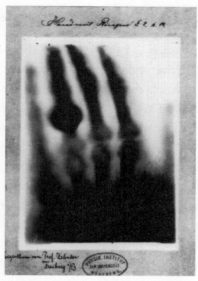

Röntgen's X-ray of his wife's hand.

lens or light, photographs taken with plate and object in darkness.' It was fascinating, baffling, and apparently useless, but physicists were intrigued and kept trying to understand what was happening.

Röntgen realised that the strange glow was some form of radiation, akin to light but invisible. He named it X-rays, where in time-honoured fashion the 'X' indicated that its nature was unknown. Apparently – we can't be sure because his notebooks haven't survived – he accidentally discovered that these rays could pass through a sheet of cardboard. This immediately made him wonder what else they could pass through. Not a thin sheet of aluminium, apparently, since only the hole showed up in the photo. Books, yes; scientific papers, yes; his wife's hand, yes. X-rays presented an unprecedented window into a living human body. Röntgen immediately grasped their medical potential, and the media were quick to publicise it. In 1896 the journal *Science* carried 23 papers on X-rays, the topic of over a thousand scientific articles that year.

It soon emerged that although X-rays did no obvious damage, repeated or long-term exposure could lead to skin burns and hair falling out. In one such instance, a child who had been shot in the head was brought to a laboratory in Vanderbilt University, and John Daniel took an X-ray with an exposure time of one hour. Three weeks later he noticed a bald patch on the child's skull where he'd placed the X-ray tube. Despite such evidence, many doctors remained convinced that X-rays were safe, blaming such damage on ultraviolet exposure or ozone, until the American radiographer Elizabeth Fleischmann died from complications caused by X-rays in 1905. The medical uses continued, but with greater caution, and better photographic plates reduced the exposure time. Today we recognise that, however useful X-rays may be, the total radiation dose must be kept to the absolute minimum. It took a while. In the 1950s, when I was about ten years old, I remember shoe shops with an X-ray machine that let you try on shoes and see how well they fitted the shape of your feet.

X-ray photographs suffered from a number of defects. They were black-and-white: black regions where the rays didn't penetrate, white where they did, and shades of grey in between. Or, more commonly, the reverse, it being simpler to make a photographic negative. Bones showed up clearly, soft tissues were largely invisible. But the most serious difficulty was that the image was two-dimensional. In effect, it flattened the internal layout, superimposing the images of all the organs between the X-ray source and the photographic plate. Of course, you could try taking more X-ray photos from other directions, but interpreting the results required skill and experience, and extra photos increased the radiation dose.

Wouldn't it be great if there were some way to image the body's interior in three dimensions?

*

As it happened, mathematicians had already made some fundamental discoveries about just that question, showing that if you take lots of two-dimensional 'flattened' images from many different directions, you can deduce the three-dimensional structure of the source of those images. They weren't motivated by X-rays and medicine, however. They were just following up on a method that had originally been invented to solve problems about waves and heat flow.

The full story has a star-studded cast, among them Galileo, who rolled balls down a slope and observed delightfully simple mathematical patterns in the distance it travelled after any given time, and Newton, who found profound patterns in the movements of the planets. Newton deduced both patterns from mathematical equations for the motion of a system of bodies acted on by forces. In his monumental *Philosophiae Naturalis Principia Mathematica* (Mathematical Principles of Natural Philosophy, usually abbreviated to *Principia*), Newton chose to explain his ideas using classical geometry, but their 'cleanest' mathematical formulation came from another of his discoveries, calculus, also found independently by Gottfried Wilhelm Leibniz. Thus reinterpreted, Newton realised that fundamental laws of nature can be expressed by differential equations – equations about the *rate* at which important quantities change over time. Thus velocity is the rate of change of position, and acceleration is the rate of change of velocity.

Galileo's patterns are at their simplest when expressed in terms of acceleration: a rolling ball moves with a constant acceleration. Its velocity therefore increases at a constant rate – it grows linearly. Its position comes from constantly increasing velocity, which implies that if it starts from rest at time zero, its position is proportional to the *square* of the elapsed time. Newton coupled this idea with another simple law, that gravity acts inversely as the square of distance, and deduced that planets travel in elliptical orbits, explaining earlier empirical deductions by Johannes Kepler.

The mathematicians of continental Europe seized on these discoveries and applied differential equations to a vast range of physical phenomena. Water waves and sound waves are governed by the wave equation, while electricity and magnetism have their own equations as well, much like the equation for gravity. Many of these equations are 'partial' differential equations, which relate rates of change in space to rates of change over time. In 1812 the French Academy of Sciences announced that its annual prize problem would be the flow of heat. Hot bodies cool down, and heat moves along materials that conduct it, which is why the metal handle of a saucepan can get quite hot as the contents cook. The Academy wanted a mathematical description of how this happens, and a partial differential equation looked plausible because the heat distribution changes in both time and space.

Joseph Fourier had sent the Academy a paper about heat flow in 1807, but they declined to publish it. The renewed challenge inspired Fourier to develop his partial differential equation for heat flow, and he won the prize. His 'heat equation' states, in mathematical form, that the heat at a given location changes over time by diffusing to neighbouring regions of space, like a drop of ink on blotting paper.

The trouble began when Fourier attempted to solve his equation, beginning with a simple case: heat in a metal bar. He noticed that there's a simple solution if the initial distribution of heat looks like a sine or cosine curve from trigonometry. Then he noticed that he could handle more complicated initial distributions by combining lots of sine and cosine curves together. He even found a calculus formula describing exactly how much each term contributes: multiply the formula for the initial heat distribution by the relevant sine or cosine, and integrate. This led him to make a bold claim: his formula, now called a Fourier series,

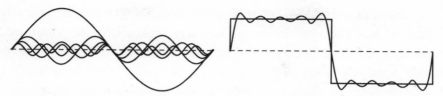

How to get a square wave from sines and cosines. *Left*: The
component sinusoidal waves. *Right*: The sum of the first
five terms of the Fourier series approximates a square wave.
Additional terms (not shown) improve the approximation.

solves the problem for *any* initial heat distribution whatsoever. In
particular, he claimed it worked for a discontinuous heat distribution, such as a square wave: one constant temperature along half
the bar, and a different one along the other half.

This claim dropped him smack in the middle of a dispute that
had been running for several decades. The same issue, indeed the
same integral formula, had already turned up in the research of
Euler and Bernoulli on the wave equation. There, the standard
'toy' example is an idealised violin string, and it's not possible
to start the string vibrating by making it discontinuous: it just
breaks. So physical intuition suggests that there might be problems
representing discontinuous functions, and mathematical intuition
backs it up by making you worry whether the trigonometric series
converges. That is, whether the sum of infinitely many sinusoidal
wiggles makes sense, and if it does, whether it adds up to the discontinuous square wave, or to something else.

Without wishing to be pejorative, part of the problem was
that Fourier was thinking like a physicist, when his critics were
thinking like mathematicians. Physically, a square wave makes
sense as a model of heat. The metal rod is idealised as a line –
which is exactly how Euler and Bernoulli idealised a violin string.
If heat is distributed uniformly along half this line, and the other
half starts out much colder, the natural model is a square wave.

Neither model is entirely accurate as a representation of reality, but the mechanics of the day was all about idealised objects such as point masses, perfectly elastic collisions, infinitely thin perfectly rigid rods, and so on. A square wave was hardly out of place in such company. Moreover, mathematically, Fourier's solution predicts that the discontinuity immediately gets smoothed by diffusion, becoming a steep but continuous curve that slowly flattens out, which makes physical sense and removes the mathematical discontinuity. Unfortunately, these arguments were too woolly to convince the mathematicians, who knew that infinite series can be subtle and troublesome. The Academy's officials reached a compromise; Fourier got the prize, but they didn't publish his memoir.

Undaunted, Fourier published his work in 1822 as *Théorie analytique de la chaleur* (The Analytical Theory of Heat). Then, to really annoy everyone, he managed to get himself appointed Secretary of the Academy, and promptly published his original prizewinning memoir, unchanged, in the Academy's journal. *Touché.*

It took about a century to resolve the mathematical issues that Fourier's claims had raised. Generally speaking, he was right about a lot of things, but wrong about several crucial ones. His method did actually work for the square wave, give or take some careful modifications about what happened exactly *at* the discontinuity. But it definitely didn't work for more complicated initial distributions. A complete understanding came only after mathematicians developed a more general notion of the integral, along with topological notions best phrased in terms of set theory.

Long before the mathematical community finally sorted out what Fourier was up to, engineers seized upon the basic idea and made it their own. They realised that the essence of his work was a mathematical transform, now called the Fourier transform, in which a complex signal that varies with time can be reinterpreted as a combination of simple signals with different frequencies.

Fourier's integral formula tells you how to change your viewpoint from the time domain to the frequency domain, and back again – which remarkably, uses almost the same formula, establishing a 'duality' between the two representations.

This duality means that you can reverse the transformation, recovering the original signal from the frequencies that it creates, like flipping a coin from heads to tails and then flipping it back again. The advantage of this procedure for engineering is that some features that are hard to detect in the time domain become obvious in the frequency domain. It can work the other way, too, so you have two very different methods for analysing the same data, and each naturally brings out particular features that the other one misses.

For instance, the response of a tall building to an earthquake looks random and chaotic in the time domain. But in the frequency domain, you might see several big spikes at particular frequencies. Those reveal the resonant frequencies at which the building responds violently to the earthquake. To design the building so that it won't fall down if an earthquake hits, you need to suppress those particular frequencies. A practical solution used in some buildings is to support the entire edifice on a concrete base, way down in the foundations, that can move sideways. Then you 'damp' this sideways motion by attaching huge weights or springs.

Another application goes back to Francis Crick and James Watson's discovery of the structure of DNA. A key piece of evidence confirming they were right was an X-ray diffraction photograph of a crystal of DNA. The technique is to pass a beam of X-rays through the crystal, which causes the rays to bend and bounce, behaviour called diffraction. The waves tend to bunch up at certain angles, governed by the diffraction law of Lawrence and William Bragg, and what shows up in a photograph is a complicated geometric arrangement of spots. This diffraction pattern is essentially a kind of Fourier transform of the positions of the atoms in the DNA molecule. Applying the inverse transform (a complicated computer calculation much easier now than it was then) you deduce the shape of the molecule. Now, as I said, the transform sometimes makes structural features obvious when they're hard to spot in the original. In this case, Crick and Watson's experience with other X-ray diffraction pictures immediately told them, without computing the inverse transform, that the molecule was some kind of helix. Other ideas refined this insight, leading to the famous double helix, later confirmed using the Fourier transform.

These are just two practical applications of the Fourier transform and its many cousins. Others include improving radio reception, removing noise caused by scratches on old vinyl records, improving the performance and sensitivity of the sonar systems used by submarines, and eliminating undesirable vibrations in cars at the design stage.

All of which, you'll notice, have nothing at all to do with the flow of heat. Unreasonable effectiveness. What matters isn't the physical interpretation of the problem – although that may well have influenced the original work – but its mathematical structure. The same methods apply to any problem with the same structure, or a similar structure, which is where scanners come into the picture.

Mathematicians also became intrigued by the Fourier transform, and they recast it in the language of functions. A function is a mathematical rule for converting a number into another number, such as 'form its square' or 'take the cube root'. All of the traditional functions such as polynomials, roots, the exponential, the logarithm, and the trigonometric functions sine, cosine, tangent, and so on, are included, but more complicated 'rules' that aren't expressed as formulas can also occur, such as the square wave that caused Fourier so much grief.

From this point of view, the Fourier transform takes a function of one type (the original signal) and transforms it into another function of a different type (the list of frequencies). There's also an inverse transform, which undoes the effect of the first one. The duality aspect, that the inverse transform is almost the same as the transform itself, is an elegant bonus. The correct context is spaces of functions with specific properties: function spaces. The Hilbert spaces used in quantum theory (Chapter 6) are function spaces, where the values of the function are complex numbers, and their mathematics is closely related to that of the Fourier transform.

All research mathematicians have acquired a very strong reflex. If someone comes up with a new idea that has remarkable and useful features, they immediately start to wonder whether there are other, similar, ideas that exploit the same trick in different circumstances. Are there other transforms like Fourier's? Other dualities? Pure mathematicians pursue such questions in their own abstract and general way, while applied ones (and engineers and physicists and Lord knows who else) start wondering how all this stuff can be used. In this case, Fourier's clever trick kick-started an entire industry of transforms and dualities, not fully mined out even today.

*

Among these variations on a theme of Fourier was one that opened the door to modern medical scanners. Its inventor was Johann Radon. He was born in 1887 in Tetschen, Bohemia, a region of Austria–Hungary, now Děčín in the Czech Republic. He was, by all accounts, friendly, personable, quiet, and scholarly. Even so, he wasn't particularly shy, and he had no trouble socialising. Like many academics and professionals, he loved music, and before radio and TV people often got together in their homes to entertain each other. Radon played the violin very well and was an excellent singer. A mathematician, he worked initially on the calculus of variations, the subject of his doctoral thesis, which led naturally into the new and fast-growing field of functional analysis. This area, initiated by Polish mathematicians led by Stefan Banach, reinterpreted key ideas of classical analysis in terms of infinite-dimensional function spaces.

In the early days of analysis, mathematicians concentrated on calculating such things as the derivative of a function, its rate of change, and its integral, the area underneath its graph. As the subject progressed, attention became focused on general properties of the operations of differentiation and integration, and how these behave for combinations of functions. If you add two functions together, what happens to their integrals? Special features of functions came to the fore: Is it continuous (no jumps)? Differentiable (smoothly varying)? Integrable (the area makes sense)? How are these properties related? How does it all work if you take a limit of a sequence of functions, or the sum of an infinite series? What *sort* of limit or sum?

Banach and his colleagues formulated these more general issues in terms of 'functionals'. Just as a function turns a number into a number, so a functional turns a function into a number or another function. Examples are 'integrate' and 'differentiate'. One great trick that the Polish mathematicians, and others, discovered is that you can take theorems about functions of numbers and turn them into theorems about functionals of functions. The resulting statement might be true, or not: the fun comes in finding out which of these happens. The idea gained traction because fairly prosaic theorems about functions turn into apparently much deeper ones about functionals, but the same simple proofs often apply. The other trick was to ignore all the technical issues about how to integrate complicated formulas in sines and logarithms and so on, and to rethink the basics. What was analysis *really* about? The most basic feature of analysis turned out to be how close together two numbers are. This is measured by their difference, in whichever order makes that difference positive. A function is continuous if small differences in the input number lead to small differences in the output. To find the derivative of a function, increase the variable by a small amount and see how the function changes in proportion to that small amount. To play similar games the next level up, with functionals, you need to define what it means for two *functions* to be close together. There are many ways to do this. You can look at the difference between their values at any given point, and make that small (for all points). You can make the integral of that difference small. Each choice leads to a different 'function space', containing all functions with specified properties, equipped with its own 'metric' or 'norm'. In the analogy with numbers and functions, the function space plays the role of the set of real or complex numbers, and a functional is a rule for turning a function in one function space into a function in another function space. The Fourier transform is a particularly important example of a functional, which turns a function into its sequence of Fourier coefficients. The inverse transform goes the other way: sequences of numbers turn into functions.

From this viewpoint, large chunks of classical analysis suddenly fitted together as examples of functional analysis. Functions of one or several real or complex variables can be thought of as rather simple functionals on rather simple spaces – the set of real numbers, the set of complex numbers, or finite-dimensional vector spaces formed by sequences of such numbers. A function of three variables is just a function(al) defined on the space of all triples of real numbers. More esoteric functionals, such as 'integrate', are defined on (say) the space of all continuous functions from three-dimensional space to the real numbers, with the metric 'integrate the square of the difference of the values of the two functions concerned'. The main difference was in the *spaces*: the real numbers and three-dimensional space are finite-dimensional, but the space of all continuous functions is infinite-dimensional. Functional analysis is just like ordinary analysis, but performed on an infinite-dimensional space.

Another major innovation of the period also fitted neatly in to this set-up: a new, more general, and more tractable theory of integration introduced by Henri Lebesgue, under the name 'measure theory'. A measure is a quantity like area or volume, which assigns a number to a set of points in some space. The new twist is that this set can be extremely complicated, though some sets are *so* complicated that even Lebesgue's concept of measure fails to apply to them.

The calculus of variations, Radon's thesis topic, positively *screams* 'functional' as soon as you observe that it's about finding functions (not numbers) with optimal properties. So it was a natural step for Radon to branch out from the classical calculus of variations into functional analysis. He did so to great effect, and several important ideas and theorems in measure theory and functional analysis are named after him.

Among them is the Radon transform, which he came across in 1917. From the viewpoint of functional analysis, it's a close mathematical cousin of the Fourier transform. Start with an image in the plane, thought of as a black-and-white picture with regions having various shades of grey. The shade can be represented by a real number from 0 (black) to 1 (white). You can squash the image flat in any direction and add together the numbers representing light and dark regions, obtaining a projection of the image. The Radon transform captures all of these squashed projections in all directions. The really important idea is the inverse transform, which lets you reconstruct the original image from these projections.

As far as I can tell, Radon studied his transform for purely mathematical reasons. His paper on the transform doesn't mention any applications; the closest it comes is a brief mention of relations to mathematical physics, specifically potential theory, which is the common ground of electricity, magnetism, and gravity. He seems much more focused on the mathematics and possible generalisations. In later work he investigated a three-dimensional analogue, in which a distribution of light and dark in space is squashed flat into all possible planes, and found a reconstruction formula for that operation too. Later, others found generalisations to higher dimensions. Radon might have been motivated by X-rays, which perform exactly this kind of projection on the distribution of organs and bones in the human body, interpreting 'light' and 'dark' as differences in opacity to X-rays. But it would take a century before his discovery found an application to devices whose abilities to probe the interior of the human body seem almost miraculous.

*

CAT (Computer Assisted Tomography) scanners – often called CT (Computed Tomography) scanners nowadays – use X-rays to create three-dimensional images of the interior of the human body. These are stored in a computer and can be manipulated to show the bones and muscles, or to locate cancerous tumours. Other kinds of scanner, such as ultrasound, are also widely used. How does a scanner find out what's inside the body without cutting it open? We all know X-rays pass easily through soft tissue, while harder tissue such as bone is more opaque to them. But an X-ray image just shows the average density of tissue when viewed from a fixed direction. How can this be transformed into a three-dimensional image? Radon opens his paper by telling us that he has solved that problem:

> When one integrates a function of two variables x, y – a *point function* $f(P)$ in the plane – subject to suitable regularity conditions along an arbitrary straight line g then one obtains in the integral values $F(g)$, a *line function*. In Part A of the present paper the problem which is solved is the inversion of this linear functional transformation, that is the following questions are answered: can every line function satisfying suitable regularity conditions be regarded as constructed in this way? If so, is f uniquely known from F and how can f be calculated?

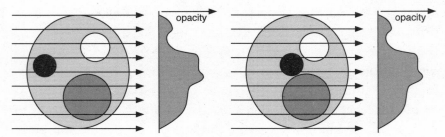

The darker a region, the more opaque it is. *Left*: Scanning a
single slice of the body from a single direction provides a graph
of the observed opacity to X-rays in that direction only. *Right*:
Different internal arrangements give the same graph.

His answer, the inverse Radon transform, is a formula that
reconstructs the internal arrangement of tissues – more precisely,
their degree of opacity to X-rays – from the entire set of projections in all directions.

To see how it works, we first describe what a single scan
(projection) of the body can observe. Such a scan is taken in one
two-dimensional slice through the body. The picture shows a schematic view of parallel X-ray beams passing through one slice of
a body that contains several internal organs of differing opacity
to X-rays. As the beam passes through these organs, the intensity
of the rays that come out the other side varies. The more opaque
the organ is, along that particular beam, the lower the intensity.
We can graph how the observed intensity varies with the position
of the beam.

In effect, a single image of this kind squashes flat the greyscale
distribution inside the body, along the direction of the beams.
Technically, this is a projection of the distribution in that direction. It's fairly clear that a single projection of this kind can't
tell us exactly how the organs are arranged. For instance, if we
move the black organ in the direction of the beam, the projection doesn't change. However, if we make another scan viewing

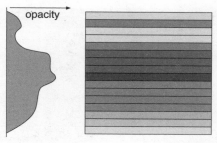

Converting the opacity graph to a series of greyscale stripes,
aligned with the direction of the X-ray beam.

the body from a vertical direction, the changed position of the
black disc has a visible effect on that opacity graph. Intuitively, we
can get even more information about the spatial positions of the
organs and tissues by making a whole series of scans, each rotated
slightly compared with the previous one, until we've viewed the
body from a large number of directions. But is that enough information to find the positions exactly?

Radon proved that if you know the opacity graphs when the
slice of the body is viewed from all possible directions, you can
deduce the two-dimensional greyscale distribution of tissues and
organs precisely. In fact, there's a very simple way to achieve this,
called backprojection. This smears the greyscale distribution out,
along the direction of the projection, but does this uniformly.
So we get a square region filled with grey stripes. The higher
the graph, the darker the corresponding stripe is. Intuitively, we
spread the grey colour uniformly along the stripe because we can't
tell from the projection where any specific internal organ is.

We can do this for every direction of the original series of
scans. Radon's inversion formula tells us that when all these
striped images are tilted to the corresponding angle and superim-
posed, so that we add up their greyscale values at each point, the
result – suitably scaled – reconstructs the original distribution of
internal organs. The next picture shows how this works when the

Left: A square. *Middle*: Backprojection from 5 directions.
Right: Backprojection from 100 directions.

original shape is a square, and we reconstruct it by backprojection from 5 and 100 different directions. The more angles we use, the better the result.

Once we've reconstructed the distribution of tissues in a single slice, we slide the body along a short distance and do the same again. And again, and again, until conceptually we've cut the body up like a sliced loaf. Then we can reassemble the slices by 'stacking' them in the computer, getting a complete description of the three-dimensional distribution of tissue. This method of detecting three-dimensional structure from a series of two-dimensional slices is known as tomography, and has long been used by microscopists to look inside solid objects such as insects or plants. The basic technique is to embed the object in wax, and then cut off very thin slices using a device like a miniature bacon-slicer, called a microtome (from the Greek *mikros* 'small' + *temnein* 'to cut'). CT scanners use the same idea, except that they do the slicing with X-rays and mathematical trickery.

After that, it's a matter of routine mathematical technique to post-process the three-dimensional data and provide all sorts of related information. We can see what the tissues would look like along a totally different slice, or show only tissues of a certain type, or colour-code muscles, organs, and bones. Whatever bells and whistles you want. The main tools here are standard image-processing methods, which ultimately rely on three-dimensional coordinate geometry.

In practice, it's not quite that simple. The scanner doesn't take infinitely many scans from a continuum of directions, just a large finite number from closely spaced discrete directions. The mathematics has to be modified to take this into account. It helps to filter the data to avoid imaging artefacts that result from using a discrete set of viewing angles. But the basic point is exactly what Radon worked out, more than fifty years before the first scanner was invented. The English electrical engineer Godfrey Hounsfield built the first working scanner in 1971. The theory had been worked out in 1956–57 by South African-born American physicist Allan Cormack, who published it in 1963–64. At the time he was unaware of Radon's results, so he worked out what he needed for himself, but later he came across Radon's paper, which is more general. Hounsfield and Cormack's development of computer assisted tomography won them the 1979 Nobel Prize in Physiology or Medicine. That machine cost $300. A commercial CT scanner today costs more like $1·5 million.

Scanners aren't just used in medicine. Egyptologists now use them routinely to find out what's inside a mummy without undoing the wrappings. They can examine the skeleton and any remaining internal organs, look for signs of fractures and various diseases, and find out where religious amulets are hidden. Museums often have exhibits of virtual mummies with a touchscreen that visitors can control, removing layers of linen bandages, then skin, then muscles, until only the bones remain. All of this relies on mathematics, embodied in a computer: three-dimensional geometry, image processing, graphical display methods.

There are many other types of scanner: ultrasound, using sound waves; Positron Emission Tomography (PET) scanners that detect subatomic particles emitted by radioactive substances injected into the body; and Magnetic Resonance Imaging (MRI), which detects magnetic effects in the nuclei of atoms, and used to be called Nuclear Magnetic Resonance (NMR) until the advertising department got worried that people might get upset by the word 'nuclear', associating it with bombs and power stations. Each type of scanner has its own mathematical story.

10

Smile, Please!

The camera's only job is to get out of the way of making photographs.

Ken Rockwell, *Your Camera Does Not Matter*

Humanity uploads about one trillion photos to the Internet every year, which suggests an over-optimistic assessment of how keen everyone else is to look at our holiday selfies, the new baby, or various other objects, some unmentionable. It's quick, it's easy, and everyone's phone is a camera. A huge amount of mathematics goes into camera design and manufacture. Those tiny high-precision lenses are technological miracles, involving some very sophisticated mathematical physics about the refraction of light by curved solids. In this chapter I want to focus on just one aspect of today's photography: image compression. Digital cameras, either stand-alone or in a phone, store very detailed images as binary files. Memory cards seem to be able to store more information than they can actually hold. So how can so many detailed pictures be contained in a small computer file?

Photographic images contain a lot of redundant information, which can be removed without loss of definition. Mathematical techniques make it possible to do this in systematic, carefully constructed ways. The JPEG standard in small point-and-click digital cameras, which until fairly recently was the most common file format and is still widely used, employs *five* separate mathematical transformations, carried out in succession. They involve discrete

Fourier analysis, algebra, and coding theory. The transformations are all built into your camera's software, which compresses the data before it's written to the memory card.

Unless, of course, you prefer RAW data – essentially, what the camera actually picked up. The capacity of a memory card is growing so fast that it's no longer essential to compress the file. But you do end up manipulating 32 MB image files when they used to be a tenth of that size, and they take longer to upload to the Cloud. Whether it's worth the hassle depends on who you are and what you want the photos for. If you're a professional, it's probably essential. If you're a point-and-click tourist like me, you can get a damned good picture of a tiger in a 2 MB JPEG file.

Image compression is a major part of the more general issue of data compression, and that remains of vital importance despite huge advances in technology. Every time the next-generation Internet gets ten times faster and has far greater capacity, some genius invents a new data format (ultra-hi definition three-dimensional video, say) that needs far more data than before, and this gets us right back to square one.

Sometimes we have no option but to squeeze every byte of capacity out of a signal channel. On 4 January 2004, on Mars, something fell from the sky, hit the ground, and bounced. In fact, Mars Exploration Rover A, otherwise known as *Spirit*, bounced 27 times, surrounded by inflatable balloons like some sort of cosmic bubble-wrap, a state-of-the-art landing. After a general check-up and various initialisation procedures, it set out to explore the surface of that alien planet, soon joined by its companion *Opportunity*. These two rovers were hugely successful, and have sent back enormous amounts of data. At the time, mathematician Philip Davis pointed out that the mission rested on a tremendous amount of mathematics, but 'the public is hardly aware of this'. Not only the public, it turned out. In 2007 Uffe Jankvist and Bjørn Toldbod, Danish postgraduate mathematicians, visited the Jet

Propulsion Laboratory at Pasadena on a journalistic mission: to uncover the hidden mathematics in the Mars Rover programme. Only to be told:

'We don't do any of that. We don't really use any abstract algebra, group theory, that kind of thing.'

This was worrying, so one of the Danes asked:

'Except in the channel coding?'

'That uses abstract algebra?'

'Reed–Solomon codes are based on Galois fields.'

'That's news to me.'

In fact, NASA's space missions use some very advanced mathematics indeed to compress data and code it in a way that corrects inevitable errors in transmission. You have to when the transmitter is about a billion kilometres from Earth and has the power of a light bulb. (Relaying data through a Mars orbiter such as Mars Odyssey or Mars Global Surveyor helps a bit.) Most of the engineers don't need to know that, so they don't. It's the public misunderstanding of mathematics in microcosm.

*

Everything on your computer, be it an e-mail, a photo, a video, or a Taylor Swift album, is stored in memory as a stream of binary digits, 'bits', 0 and 1. Eight bits make a byte, and 1,048,576 bytes make a megabyte (MB). A typical low-resolution photo occupies about 2 MB. Although all digital data take this form, different applications use different formats, so the meaning of the data depends on the application. Each type of data has hidden mathematical structure, and processing convenience is often more important than file size. Formatting data conveniently can make it redundant – using more bits than the actual information content requires. This provides an opportunity for data compression by removing redundancy.

Written (and spoken) English is highly redundant. As proof,

here's a phrase from earlier in this chapter with every fifth character deleted:

surr_unde_ by _nfla_able ball_ons _ike _ome _ort _f co_mic

You can probably work out what it says, with very little thought or effort. The information that remains is enough to reconstruct the whole of the original phrase.

Nonetheless, you'll also find this book much easier on the eye if I don't persuade my publisher to save ink by deleting every fifth character. Proper words are easier for the brain to process, because that's what it's learned to do. However, when you want to transmit a string of bits to someone else, rather than process those data using some app, a shorter sequence of 0's and 1's is more efficient. In the early days of information theory, pioneers like Claude Shannon realised that redundancy makes it possible to code a signal using fewer bits. In fact, he proved a formula stating how much shorter a code can make a signal, for a given amount of redundancy.

Redundancy is essential, because messages that aren't redundant can't be compressed without losing information. The proof is a simple counting argument. Suppose, for instance, we're interested in messages that are ten bits long, like 1001110101. There are precisely 1024 such bit strings. Suppose we want to compress ten bits of data into an 8-bit string. There are precisely 256 such strings. So we have four times as many messages as there are compressed strings. There's no way to assign an 8-bit string to every 10-bit string so that different 10-bit strings get different 8-bit ones. If every 10-bit string can occur with equal probability, it turns out there's no clever way round this limitation. However, if some 10-bit strings are very common, and others are very uncommon, we can choose a code that assigns short bit strings (say 6 bits) to the commonest messages, and longer ones (maybe 12 bits) to the uncommon ones. There are oodles of 12-bit strings, so we won't run out. Each time one occurs it adds two bits to the length, but every time a common message occurs it deducts four bits. With suitable probabilities, more bits get removed than are added.

A whole branch of mathematics, coding theory, has grown up around such techniques. They're generally far subtler than the one I've just outlined, and they usually exploit features of abstract algebra to define the codes. This shouldn't be too surprising: we saw in Chapter 5 that, at heart, codes are mathematical functions and that number-theoretic functions are particularly useful. The aim there was secrecy, whereas here it's data compression, but the same general point applies. Algebra is about *structure*, and so is redundancy.

Data compression, hence also image compression, exploits redundancy to create codes that shorten data of some specific type. Sometimes the compression method is 'lossless': the original information can be reconstructed exactly from the compressed version. Sometimes data are lost, and the reconstruction is only an approximation to the original data. That would be bad for, say, bank balances, but it's often fine for images: the trick is to set things up so that the approximation still looks like the original image to the human eye. Then the information that gets irretrievably lost didn't matter much to begin with.

Most real-world images are redundant. Holiday snaps often contain big blocks of blue sky, often pretty much the same shade of blue, so a lot of pixels all containing the same number could be replaced by two pairs of coordinates for the opposite corners of a rectangle, and a short code meaning 'colour *this* region *that* shade of blue'. This method is lossless. It's not what's actually used, but it illustrates why lossless compression is possible.

*

I'm old-fashioned; that is, I use camera technology that's, *ooooh*, about ten years old. Disgraceful! I'm sufficiently tech-savvy to use my phone as a camera, sometimes, but it doesn't come as a reflex, and on major holiday trips like a tiger safari in India's national parks, I take a small point-and-click digital camera. It creates image files with names like IMG_0209.JPG. The JPG identifier shows that the file format is JPEG, the initials of the Joint Photographic Experts Group, and it indicates a system of data compression. JPEG is an industry standard, though it's evolved over the years and now comes in several technically different forms.

The JPEG format[59] uses at least five different steps in turn, most of which compress the data from the previous step (the original raw data for step one). The others recode it for further compression. Digital images are composed of millions of tiny squares, called pixels – picture elements. Raw camera data assigns bit strings to each pixel to represent both colour and brightness. Both quantities are represented simultaneously as proportions of three components: red, green, and blue. Low proportions for all three correspond to pale colours, high proportions to dark. These numbers are converted into three related ones that better correspond to how the human brain perceives images. The first, luminance, gives the overall brightness, and it's measured by numbers ranging from black, through increasingly light shades of grey, to white. If you stripped out the colour information you'd be left with an old-fashioned black-and-white image – actually, many shades of grey. The other two numbers, known as chrominance, are the differences between this and the amounts of blue and red light, respectively.

Symbolically, if R = red, G = green, and B = blue, then the initial numbers for R, G, and B are replaced by luminance $R + G + B$ and two chrominances $(R + G + B) - B = R + G$ and $(R + G + B) - R = G + B$. If you know $R + G + B$, $R + G$, and $G + B$ you can calculate R, G, and B, so this step is lossless.

Step two isn't lossless. It cuts the chrominance data down to smaller values by coarsening the resolution. This step alone halves the size of the data file. It's acceptable because, compared to what the camera 'sees', the human visual system is more sensitive to brightness, and less sensitive to colour differences.

Step three is the most mathematical. It compresses the luminance information, using a digital version of the Fourier transform, which we met in Chapter 9 in connection with medical scanners. There the original Fourier transform, which converts signals to their component frequencies or vice versa, was modified to represent projections of greyscale images. This time we represent the greyscale images themselves, but in a simple digital format. The image is split into tiny 8 × 8 blocks of pixels, so there are there are 64 different possible luminance values, one for each pixel. The discrete cosine transform, a digital version of the Fourier transform, represents this 8 × 8 greyscale image as a superposition of multiples of 64 standard images (overleaf). Those multiples are the *amplitudes* of the corresponding images. These images look like stripes and chequerboards of various widths. Any 8 × 8 block of pixels can be obtained in this manner, so again this step is lossless. In coordinates on the block, these standard blocks are discrete versions of $\cos mx \cos ny$ for various integers m and n, where x runs horizontally and y vertically and both run from 0 to 7.

Although the discrete Fourier transform is lossless, it's not pointless, because it makes step four possible. This step again relies on a lack of sensitivity in human vision, which creates redundancy. If brightness or colour vary over large regions of an image, we notice. If they vary within small regions, the visual system smooths them out and we see only the average. This is why printed images are comprehensible, even though on close examination they represent shades of grey by patterns of black dots on the white background of the paper. This feature of human vision means that patterns with very fine stripes are less important, so their amplitudes can be recorded with smaller precision.

Step five is a technical trick, called a 'Huffman code', to record the amplitudes of the 64 basic patterns more efficiently. David Huffman invented this method in 1951 when still a student. He'd

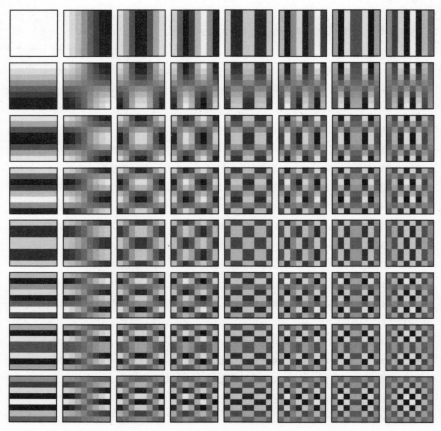

The 64 basic patterns of the discrete cosine transform.

been set the problem of writing a term paper on optimally efficient binary codes, but was unable to prove that any existing code was optimal. About to give up, he thought of a new method and then proved it was best possible. Roughly speaking, the problem is to encode a set of symbols using binary strings, and then use that as a dictionary to convert a message into a coded form. This must be done so that the total length of the coded message is minimised.

For example, the symbols might be the letters of the alphabet. There are 26 of these, so you could just assign a 5-bit string, say

A = 00001, B = 00010, and so on. You need five bits because four bits gives only 16 strings. But this would be inefficient, because letters that occur infrequently, such as Z, use the same number of bits as common ones like E. It would be better to assign a short string, such as 0 or 1, to E, and gradually longer strings as the letters become less probable. However, because the code strings have different lengths, you need extra information to tell the recipient where to split the strings up into separate letters. This can be done by recognising a prefix at the front of the code string, but the code must be prefix-free: no code string appears as the start of some longer code string. If it did, you wouldn't know where that code string ends. An uncommon letter like Z now needs many more bits, but because it's uncommon, the shorter strings for E more than compensate. The overall length of a typical message is shorter.

Huffman codes achieve this goal by forming a 'tree', a kind of graph that has no closed loops, and is very common in computer science because it represents an entire strategy of yes/no decisions, each depending on the previous one. The leaves of the tree are the symbols A, B, C, ..., and two branches emerge from each leaf, corresponding to the two bits 0 and 1. Each leaf is labelled with a number, called its weight, showing how frequently the corresponding symbol occurs. The tree is constructed step by step by merging the two least common leaves into a new 'mother' leaf, while they become 'daughter' leaves. The weight assigned to the mother leaf is the sum of the weights of the two daughters. This process continues until all symbols have been merged in this manner. Then the code string for a symbol is read off from the path that leads to that symbol.

For example, the picture shows at top left five symbols A, B, C, D, E, and numbers 18, 9, 7, 4, 3 indicating how common they are. The two least common symbols are D and E. The second stage, top middle, merges these to form a mother leaf (unnumbered) with weight $4 + 3 = 7$, and the symbols D and E become daughters. The

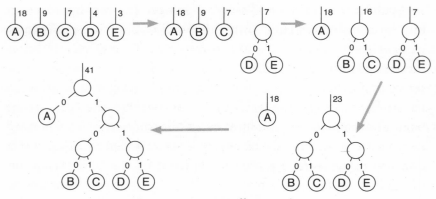

Constructing a Huffman code.

two branches leading to them are labelled 0 and 1. This process is repeated until all symbols have been merged (bottom left). Now we read off the code strings by following paths down the tree. A is reached by a single branch labelled 0. B is reached by the path 100, C by the path 101, D by 110, and E by 111. Notice how A, the most common symbol, gets a short path, while less common symbols get longer paths. If instead we'd use a fixed-length code, it would need at least three bits to give five symbols, because there are only four 2-bit strings. Here the longest strings are 3-bit, but the commonest is 1-bit, so on average this code is more efficient. This procedure ensures that the code is prefix-free, because any path that leads to a symbol, stops at that symbol. It can't continue to another symbol. Moreover, by starting with the least common symbols, the most common symbols are assigned the shortest paths. It's a very clever idea, easy to program, and conceptually very simple once you've figured it out.

When your camera creates a JPEG file, its built-in electronics does all these sums on the fly as soon as you've taken a picture. The compression process isn't lossless, but most of us would never notice; in any case, our computer screens or paper printouts don't get colours and brightness exactly right anyway, unless they've been carefully calibrated. Direct comparison of the original image and the compressed version makes the differences more obvious, but even then it takes an expert to spot the difference when the file size has been reduced to 10% of the original. Ordinary mortals notice only when the reduction gets down to about 3%. So JPEG can store ten times as many images on a given memory card as the original RAW data. This complicated five-step procedure, carried out in a flash behind the scenes, is how the magic works, and it uses at least five different areas of mathematics.

*

Another way to compress images emerged in the late 1980s from fractal geometry. A fractal, you'll recall, is a geometric shape with detailed structure on all scales, like coastlines and clouds. Associated with any fractal is a number, called its fractal dimension, which is a measure of how rough or wiggly the fractal is. Typically, the fractal dimension isn't a whole number. A useful class of mathematically tractable fractals comprises those that are self-similar: small pieces of them, suitably magnified, look just like bigger pieces of the whole. The classic example is a fern, which is composed of dozens of smaller fronds, each of which looks like a miniature fern. Self-similar fractals can be represented by a mathematical set-up called an iterated function system (IFS). This is a set of rules telling you how to shrink copies of the shape and move the resulting tiles so that they fit together to give the whole thing. You can reconstruct the fractal from these rules, and there's even a formula for the fractal dimension.

In 1987 Michael Barnsley, a mathematician fascinated by fractals, realised that this property might form the basis of an image compression method. Instead of using vast amounts of data to encode every tiny detail of the fern, you just encode the corresponding IFS, which requires much less data. Software can reconstruct the image of the fern from the IFS. Together with Alan Sloan he

A fractal fern, made from three transformed copies of itself.

formed a company, Iterated Systems Inc., which was granted over 20 patents. In 1992 the company made a breakthrough: an automated method for finding suitable IFS rules, which searches for small regions of the image that can be seen as shrunken versions of slightly larger regions. So it uses a lot more tiles to cover the image. However, it's also completely general, applicable to any image, not just ones that are obviously self-similar. Fractal image compression didn't reach the level of success of JPEG, for various reasons, but it was used in several practical applications. The most successful was probably Microsoft's digital encylopaedia *Encarta*, where all of the main images were compressed using an IFS.

Throughout the 1990s the company made strenuous attempts to extend the method to video compression, but none took off, mainly because computers weren't fast enough and didn't have sufficient memory then. It took 15 hours to compress one minute of video. All that has now changed, and fractal video compression ratios of 200:1 have been achieved at about one minute per

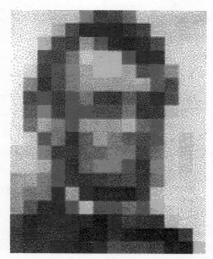

Who's this? Squint.

video frame. However, improved computing power also makes other methods feasible, and fractal video compression has been abandoned for the moment. But the underlying idea was useful for a time, and it remains an intriguing possibility.

*

Humans have a very strange trick for deciphering a poor image: we screw up our eyes and squint. It's amazing how often this helps us work out what the picture actually is, especially if it's a bit blurred, or if it's a computer image with very coarse pixels. There's a famous picture composed of 270 black, white, and grey squares, which Leon Harmon at Bell Labs created in 1973 for an article on human perception and computer pattern recognition. Who is it? It eventually becomes vaguely recognisable as Abraham Lincoln if you stare at it, but if you squint, it really *looks* like Lincoln.

We all do this, so we know it works, but it seems crazy. How can you improve a bad picture by making your vision worse? Part of the answer is psychological: by squinting, we put our brain's visual processing system into 'poor image mode', which presumably triggers special image-processing algorithms that evolved to handle poor data. But another part is that, paradoxically, squinting acts as a kind of preprocessing step, which cleans the image up in certain useful ways. For example, it blurs Lincoln's pixellated boundaries so that he no longer looks like a stack of grey building blocks.

About forty years ago, mathematicians started to investigate a precise and versatile equivalent of the human squint, called wavelet analysis. The technique applies to numerical data as well as visual images, and it was originally introduced in order to extract structure at some particular spatial scale. Wavelets let you detect the wood while remaining unaware that it's composed of lots of complicated trees and bushes.

The original impetus was largely theoretical: wavelets were great for testing scientific theories of things like turbulent fluid flow. More recently, wavelets acquired some extremely down-to-earth applications. In the United States the Federal Bureau of Investigation (FBI) applies wavelets to store fingerprint data more cheaply, and law enforcement agencies in other countries have followed their lead. Wavelets don't just let you analyse images: they let you compress them.

In JPEG, images are compressed by discarding information that's less relevant to human vision. However, information is seldom represented in a way that makes it obvious which bits are less relevant. Suppose you want to e-mail to a friend a drawing on a rather dirty piece of paper. There are lots of little black specks as well as the drawing itself. You or I can look at the picture and see immediately that the specks are irrelevant, but a scanner can't. It just scans the page line by line, representing the image as a long string of binary black/white signals, and it can't tell whether any particular black dot is an essential part of the drawing or an irrelevant speck. Some 'specks' might actually be the eyeball of a distant cow or spots on a cartoon leopard.

The main obstacle is that the scanner's signals don't represent the image data in a way that makes it easy to recognise and remove unwanted items. However, there are other ways to represent data. The Fourier transform replaces a curve by a list of amplitudes and frequencies, encoding the same information in a different way. And when data are represented in different ways, operations that are difficult or impossible in one representation may become easy in the other. For example, you can start with a telephone conversation, form its Fourier transform, and strip out all parts of the signal whose Fourier components have frequencies too high or too low for the human ear to hear. Then you can reverse transform the result to get sounds that, to human hearing, are identical to the original. Now you can send more conversations over the same communication channel. You can't play this game directly on the original untransformed signal, because that doesn't have 'frequency' as an obvious characteristic.

For some purposes, Fourier's technique has one fault: the component sines and cosines go on forever. The Fourier transform does a poor job of representing a compact signal. A single 'blip' is a simple signal, but it takes hundreds of sines and cosines to produce a moderately convincing blip. The problem isn't getting the shape of the blip right, but making everything outside the blip equal to zero. You have to kill off the infinitely long wavy tails of all those sines and cosines, which you do by adding even more high-frequency sines and cosines in a desperate effort to cancel out the unwanted junk. Ultimately, the transformed version becomes more complicated, and needs more data, than the original blip.

The wavelet transform changes all that by using blips as its basic components. This isn't easy, and you can't do it with any old blip, but it's clear to a mathematician how to get started. Choose some particular shape of blip to act as a mother wavelet. Generate daughter wavelets (and granddaughters, great-granddaughters,

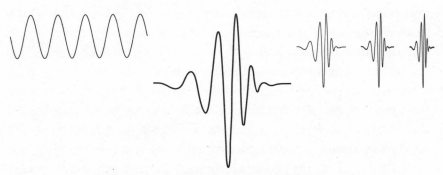

Left: A sine curve goes on forever. *Middle*: A wavelet
is localised. *Right*: Three more generations.

whatever) by sliding the mother wavelet sideways into various
positions, and expanding her or compressing her by a change of
scale. To represent a more general function, we add up suitable
multiples of these wavelets on different scales. In the same way,
Fourier's basic sine and cosine curves are 'mother sinelets', and all
the other frequency sines and cosines are daughters.

Wavelets are designed to describe bliplike data efficiently.
Moreover, because the daughter and granddaughter wavelets are
just rescaled versions of mother, it's possible to focus on particular levels of detail. If you want to delete small-scale structure, you
remove all the great-granddaughter wavelets in the wavelet transform. Imagine transforming a leopard into wavelets – a few big
ones for the body, smaller ones for the eyes, nose, and spots, then
tiny ones for hairs. To compress the data but keep it looking like a
leopard, you decide that individual hairs don't matter, and remove
those great-granddaughter wavelets. The spots remain, and it still
looks like a leopard. You can't do this anything like as easily – if
at all – using a Fourier transform.

Most of the mathematical tools required to develop wavelets had been around in an abstract form for half a century or
more, in Banach's area of functional analysis. When wavelets hit
the streets, it transpired that the esoteric machinery of functional

analysis was exactly what was needed to understand them and develop them into an effective technique. The main prerequisite for the functional-analytic machine to roar into life was a good shape for the mother wavelet. We want all the daughter wavelets to be mathematically independent of mother, with no overlap in the information encoded by mother and daughter, and no part of any daughter being redundant. In functional-analytic terminology, mother and daughter must be orthogonal.

In the early 1980s geophysicist Jean Morlet and mathematical physicist Alexander Grossmann came up with a feasible mother wavelet. In 1985 the mathematician Yves Meyer improved Morlet and Grossmann's wavelets. The discovery that blew the whole field wide open was made in 1987 by Ingrid Daubechies. Previous mother wavelets looked suitably bliplike, but they all had a very tiny mathematical tail that wiggled off to infinity. Daubechies constructed a mother wavelet with no tail at all: outside some interval, mother is always *exactly* zero. Her mother wavelet was a genuine blip, confined entirely to a finite region of space.

*

Wavelets act as a kind of numerical zoom lens, focusing on features of the data that occupy particular spatial scales. This ability can be used to analyse data, but also to compress it. By manipulating the wavelet transform, the computer 'squints' at the image and discards unwanted scales of resolution. This is what the FBI decided to do in 1993. At that time, the Bureau's fingerprint database contained 200 million records, stored as inked impressions on paper cards, and the records were being modernised by digitising the images and storing the results on a computer. An obvious advantage is being able to search rapidly for prints that match those found at the crime scene.

A conventional image with sufficient resolution creates a computer file ten megabytes long for each fingerprint card. The FBI's archive occupies 2,000 terabytes of memory. At least 30,000 new cards are received every day, so the storage requirement grows by 2·4 trillion binary digits every day. The FBI desperately needed data compression. They tried JPEG, but this becomes useless for fingerprints (unlike holiday snaps) when the 'compression ratio' – the ratio of the size of the original data to that of the compressed data – becomes high, around 10:1. Then the uncompressed images become unsatisfactory because of 'tiling artefacts' in which the sub-division into 8 × 8 blocks leaves marked boundaries. Naturally, the method wasn't much use to the FBI unless it could deliver compression ratios of at least 10:1. Tiling artefacts aren't just an aesthetic problem: they seriously impair the ability of algorithms to search for matching fingerprints. Alternative Fourier-based methods also introduce objectionable artefacts, all of which can be traced back to the problem of infinite 'tails' in Fourier sines and cosines. So Tom Hopper at the FBI, and Jonathan Bradley and Chris Brislawn from the Los Alamos National Laboratory, decided to encode digitised fingerprint records with wavelets, using a method known as wavelet/scalar quantisation, or WSQ for short.

Instead of removing redundant information by creating tiling artefacts, WSQ removes fine detail throughout the image – detail so fine as to be irrelevant to the eye's ability to recognise the structure of the fingerprint. In the FBI's trials, three different wavelet methods all outperformed two Fourier methods such as JPEG. On balance WSQ emerged as the most sensible method. It provides a compression ratio of at least 15:1, reducing the cost of storage memory by 93%. WSQ is now a standard for the exchange and storage of fingerprint images. Most American law enforcement agencies use it for compressed fingerprint images at 500 pixels per inch. For higher-resolution fingerprints they use JPEG.[60]

Wavelets are turning up almost everywhere. Dennis Healy's team has applied wavelet-based image-enhancing methods to CT, PET, and MRI scans. They've also used wavelets to improve the strategies by which the scanners acquire their data in the first

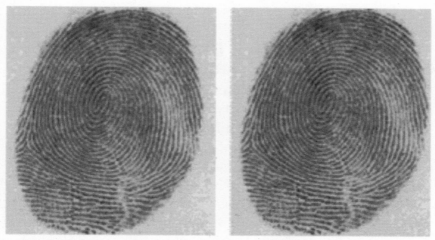

Fingerprints. *Left*: Original. *Right*: After compression to 1/26 data size.

place. Ronald Coifman and Victor Wickerhauser have used them to remove unwanted noise from recordings. One triumph was a performance of Johannes Brahms playing one of his own *Hungarian Dances*, originally recorded on a wax cylinder in 1889, which partially melted. It was re-recorded on a 78 rpm disc. Coifman started from a radio broadcast of the disc, by which time the music was virtually inaudible amid the surrounding noise. After wavelet cleansing, you could hear what Brahms was playing – not perfectly, but you *could* hear it.

Forty years ago functional analysis was just another arcane area of abstract mathematics whose main applications were to theoretical physics. The arrival of the wavelet has changed all that. Functional analysis now provides the underpinnings needed to develop new types of wavelet with special features that make them important in applied science and technology. Wavelets are having an unseen impact on all of our lives today – in crime prevention, in medicine, in the next generation of digital music. Tomorrow they will take over the world.

11

Are We Nearly There Yet?

A journey of a thousand miles begins with a single step.
                                    Lao-tzu, *The Way of Lao-tzu*

Every parent who drives knows the scenario. The family is heading out to visit grandma, three hundred miles and six hours away. The kids are in the back. Half an hour into the journey comes the plaintive cry: 'Are we nearly there yet?'

I have a bone to pick with my transatlantic cousins, who seem to be convinced that the phrase is 'Are we there yet?' No doubt it is, in the USA, but it shouldn't be, because this variant clearly results from a misunderstanding. The answer to the second version is always obvious: either we *are* there, and it's superfluous, or we're not, and it's pointless to ask. No, what happened was that on any long journey, when the kids got fractious, the kindly (or possibly just annoyed) parent reassured them. 'Nearly there now.' Even if there were five hours still to go. It shut them up for a while. Anyway, after several trips, the kids started dropping gentle hints in despair rather than hope: 'Are we *nearly* there yet?' This is a sensible question, because you can't tell by looking out of the window. Unless you know the landmarks, of course. We had a cat that did.

Are we nearly there yet? Where *are* we? Two decades ago you needed a map, good map-reading skills, and a navigator in the passenger seat to find out. Today, it's all been outsourced to electronic wizardry. You look at the satnav. True, people sometimes end up in the middle of a field. One car drove into a river recently, guided there by courtesy of satnav. You need to look at the road as well. But even that can go wrong. Last year we ended up in the grounds of a country house when we were looking for a Bed and Breakfast, because the satnav couldn't resolve the difference between a real road that looked like the driveway of a house, and the driveway of a house that looked like a real road.

Satellite navigation looks like magic. You have a screen in the car, displaying part of a map. The map shows exactly where you are. As you drive along, the map moves so that the symbol for your car is always in the right place. The device knows in which direction you're heading, and the name or number of the road you're currently on. It alerts you to traffic hold-ups. It knows where you're going, how fast you're going, when you're over the speed limit, where the traffic cameras are, and how long before you get there. Train the kids to read that, and they need never ask again.

'Any sufficiently advanced technology,' the great science fiction author and futurologist Arthur C. Clarke wrote, 'is indistinguishable from magic.' Another SF writer, Gregory Benford, recast that as: 'Any technology distinguishable from magic is insufficiently advanced.' Satnav is sufficiently advanced, but it's not magic. How does it work?

It knows where you're going because you told it. You touched letters and numbers on the screen. That bit's obvious. It's also the *only* bit that's obvious. The remaining magic rests on high technology – orbiting satellites, a lot of them; radio signals; codes; pseudorandom numbers; a lot of clever computer processing. Algorithms to find the quickest/cheapest/least environmentally damaging route. Fundamental physics is vital: orbital mechanics based on Newton's law of gravity, augmented by Einstein's special theory of relativity *and* his general theory, which improves on Newton. Out in space the satellites whirl, transmitting timing signals; at your end, nearly everything happens in one tiny computer chip. Plus some memory chips to store the map and so on.

We don't see any of these things, so we see magic instead.

Needless to say, much of the magic is mathematical, requiring liberal doses of mathematics of many kinds, not to mention vast amounts of physics, chemistry, materials science, and engineering. Psychiatric treatment might also be a good idea for some users, but hey.

Ignoring the manufacture and design of satellites, and the technology required to get them up in space, satellite navigation still involves at least seven areas of mathematics, and wouldn't work without them. The ones I have in mind are:

- Calculating the trajectories of the launch rockets to put the satellites into orbit
- Designing a set of orbits that gives good coverage: at least three, and preferably more, satellites need to be visible from any given location at any time
- Using a pseudorandom number generator to create the signals, making it possible to measure very accurately how far away each satellite is
- Using trigonometry and orbital data to deduce the location of your car
- Using the equations of special relativity to correct the calculations for the effect of the high speeds of the satellites on the passage of time
- Using the equations of general relativity to correct the calculations for the effect of the Earth's gravity on the passage of time
- Solving a variant of the Travelling Salesperson Problem to find the best route by whichever criterion you selected: fast, short, environmentally friendly.

I'll discuss most of these in more detail over the next few pages, concentrating on the more surprising ones.

*

Satnav relies on extraordinarily precise timing signals produced by highly accurate atomic clocks, sent from a number of special orbiting satellites. A caesium clock, left to its own devices, is accurate to 5 parts in 10^{14}, or 4 nanoseconds per day. That corresponds to an error in your position of about one metre per day. To compensate for this gradual drift, the clocks are periodically reset from the ground station. There are other sources of timing errors, which I'll come back to.

There are several satellite navigation systems now, but I'll concentrate on the first and most widely used, the Global Positioning System (GPS). The project began in 1973, under the aegis of the US Department of Defense. The core of the system is a set of orbital satellites: originally 24 of them, now 31. The first prototype satellite was launched in 1978, and the complete set became operational in 1993. Initially GPS was restricted to military uses, but President Ronald Reagan's executive order of 1983 made it available to civilians in a lower-resolution form. GPS is in the process of upgrading, and several nations now operate their own satellite positioning systems, starting with Russia's Global Navigation Satellite System (GLONASS), accurate to within two metres. In 2018 China began its BeiDou Navigation Satellite System, which should be up and running any time now. The European Union's system is called Galileo. Britain has now left the EU and will not take part in Galileo, but in a triumph of ideology over common sense the UK government has announced that Britain will develop and launch its own system. India is setting up NavIC, and Japan is constructing the Quasi-Zenith Satellite System (QZSS), which will eliminate reliance on GPS by 2023.

Operationally, GPS comprises three 'segments': space (the satellites), control (ground stations), and user (you in your car). The satellites send out timing signals. The control segment monitors the orbits of the satellites and the accuracy of their clocks, and if

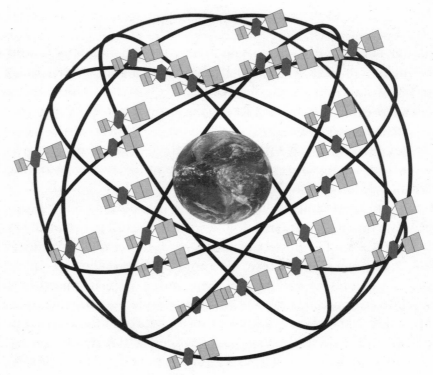

Original GPS constellation of 24 satellites, four
of them in each of six distinct orbits.

necessary transmits instructions to modify the orbit and reset the clock. The user has a cheap low-powered receiver small enough to fit into a mobile phone, to tell apps where they are.

The set of satellites is generally called a 'constellation', the long-established name for an arrangement of stars in the night sky. The original GPS constellation comprises 24 satellites, each in a roughly circular orbit 20,200 km (12,600 miles) above the Earth, or 26,600 km (16,500 miles) from its centre. I'll ignore later extra satellites, which don't affect the main ideas; they make the system more reliable and more accurate. There are six orbits, in planes that meet the equator at angles of 55°, spaced equally around the equator. Each orbit is occupied by four equally spaced satellites, perpetually chasing each other's tails. The radius of the orbit is chosen, using the mathematics of orbits, so that the satellite returns to the same position in its orbit every 11 hours 58 minutes. This ensures that it's above almost the same location on the Earth twice per day, but slowly drifts.

The next mathematical feature is the geometry of the orbits. This configuration of satellites and orbits means that at any given time at least six of the satellites are visible (that is, signals from them can be received) from any point on the planet. Which six depends on where you are, and this set changes as time passes, because the Earth spins and the satellites revolve in their orbits.

GPS is designed so that users don't need to transmit any information to the satellites. Instead, they have a receiver that picks up timing signals from whichever satellites are visible. The receiver processes the timing data to work out where it is. The basic principle is simple, so let's look at that first. Then I'll point out some of the tweaks that are needed to make it work in the real world.

Start with one satellite. It sends you timing signals, from which your receiver works out how far away the satellite is at that instant. (We'll see later how that deduction is made.) Perhaps this distance is 21,000 km. This information places you on the surface of a sphere, centred on the satellite and of radius 21,000 km. That's not terribly useful on its own, but at least five other satellites are visible at the same instant. Let me refer to them as satellite 2, satellite 3, and so on, up to satellite 6. Each transmits signals, which you receive simultaneously, and each signal places you on another sphere, also centred at the satellite: spheres 2, 3, 4, 5, 6. The signal from satellite 2, combined with that from satellite 1, places you on the intersection of spheres 1 and 2, which is a circle. Satellite 3 contributes another sphere, which meets sphere 1 in another circle. The two circles cross at two points, each of which lies on all three spheres. The signal from satellite 4 provides sphere 4, which generally distinguishes which of the two points is your correct location.

In a perfect world, we could stop there, and satellites 5 and 6 would be superfluous. In reality, it's not that straightforward. Everything is subject to errors. The Earth's atmosphere can degrade the signal, there might be electrical interference, whatever. For a start, this implies that your location is *near* the sphere concerned, rather than *on* it. Instead of the surface of the sphere, your position lies within a thickened-up shell that contains the surface. So four satellites and four signals can pin you down to some level of accuracy, but not perfectly. To do better, GPS uses the extra satellites. Their thickened-up spheres cut the region down even more. At this stage the equations determining your location are almost certainly inconsistent if you ignore likely errors, but borrowing an old trick from statistics, you can work out the best estimate of your position by minimising the total error. This is called the Method of Least Squares, and was introduced by Gauss in 1795.

The upshot is that your GPS receiver just has to do a systematic series of relatively simple geometric calculations, which leads to the best estimate it can make of your location. By comparing that with the detailed shape of the Earth, it can even work out how high you are relative to sea level. Heights are generally less accurate than latitude/longitude positions.

*

'Send timing signals' sounds simple, but it's not. If you hear a clap of thunder, you know there's a storm around, but the clap alone doesn't tell you how far away it is. If you also see the lightning, which arrives before the thunderclap because light travels faster than sound, you can use the time difference between the two signals to estimate how far away the lightning was – the rule of thumb is five seconds per mile. However, the speed of sound depends on the state of the atmosphere, so this rule isn't totally accurate.

GPS can't use sound waves as a second signal, for obvious

reasons – too slow, and space is a vacuum so sound can't travel anyway. But the underlying idea, that you infer a time difference by comparing two distinct but related signals, points in the right direction. Each satellite sends a sequence of 0/1 pulses that doesn't contain repetitions, unless you wait a very long time for the entire sequence to repeat. The GPS receiver can compare the string of 0's and 1's that it gets from the satellite with the same string produced by a local source. The satellite signal is delayed because it has to cover the distance between the satellite and the receiver, and you can deduce the time delay by aligning the signals and seeing how far one has to be displaced to match the other.

We can illustrate this process using words from this book rather than 0's and 1's.

Suppose that the signal received from the satellite is:

aligning the signals and seeing how far one

while, at the same time, the reference signal from just down the road is:

seeing how far one has to be displaced

Then we can slide the local signal along until the words agree, like this:

aligning the signals and **seeing how far one**
                      seeing how far one has to be displaced

Now we can see that the satellite signal is arriving four words later than the local signal.

All that remains now is to generate suitable bit strings. A simple way to generate strings of 0's and 1's with very infrequent repetitions is to toss a coin millions of times and record 0 for a head and 1 for a tail. Each bit occurs with probability

1/2, so a particular string of, say, 50 bits occurs with probability $1/2^{50}$, which is about one chance in a quadrillion. On average, it will repeat about a quadrillion steps along the string. When you compare such a signal with a version that's been displaced by a much shorter amount, the 'correct' displacement, giving the best match, is unique.

Computers, however, aren't good at tossing coins. They follow specific instructions, and the whole point is that they should do that precisely and without error. Fortunately there are precise mathematical processes that can generate strings of bits that *look* random, in any reasonable statistical sense, even though the actual procedure is deterministic. Such a method is known as a pseudo-random number generator. This is the third major mathematical ingredient in GPS.

In practice, the bit stream from the pseudorandom number generator is combined with other data that GPS requires, a technique called modulation. The satellite broadcasts its data at a relatively slow rate, 50 bits per second. It combines this signal with a much faster bit stream from the pseudorandom number generator, at a rate of over a million *chips* per second. A chip is much like a bit but taking values $+1$ or -1 rather than 0 or 1. Physically it's a square wave pulse with amplitude either $+1$ or -1. 'Modulation' means that the original data string is multiplied by the value of the chip at each instant. Because the other data change very slowly in comparison, the 'slide and match' technique still works well enough, but sometimes the match is identical, and sometimes one signal is minus the other. Using the statistical method of correlation, you just have to slide the signals until the correlation is sufficiently high.

In fact, GPS does the same thing again with another pseudo-random number, modulating the signal at a rate ten times faster. The slower one is called Coarse Acquisition Code, and is for civilian use. The faster one, Precise Code, is reserved for the military. It's also encrypted, and it takes seven days to repeat itself.

Pseudorandom number generators are generally based on abstract algebra, such as polynomials over finite fields, or number theory, such as integers to some modulus. A simple example of the latter is a linear congruential generator. Choose some modulus m, two numbers a and b (mod m), and a starting number x_1 (mod m). Then define successive numbers x_2, x_3, x_4, and so on, by the formula

$$x_{n+1} = ax_n + b \quad (\text{mod } m)$$

The effect of a is to multiply the current number x_n by a constant factor a, and then b shifts that value along by a fixed amount. This gives the next number in the sequence; repeat. For instance, if $m = 17$, $a = 3$, $b = 5$, and $x_1 = 1$, then we get the sequence

$$1 \ 8 \ 12 \ 7 \ 9 \ 15 \ 16 \ 2 \ 11 \ 4 \ 0 \ 5 \ 3 \ 14 \ 13 \ 10$$

which then repeats indefinitely. This has no obvious patterns to the eye. In practice, of course, we make m much bigger. There are some mathematical conditions that ensure that the sequence takes a long time to repeat, and that it satisfies reasonable statistical tests for randomness. For example, after turning the output into binary, every number (mod m) should appear equally often, on average; so should every string of 0's and 1's of given length, up to some reasonable size.

Linear congruential generators are too simple to be secure, and more complex variants have been devised. An example is the Mersenne twister, invented in 1997 by Makoto Matsumoto. Many of you will have this, since it's used in dozens of standard software packages, among them Microsoft Excel spreadsheets. The Mersenne twister combines prime numbers, which make the mathematics easier, and nice binary expressions, which make the computations easier. A Mersenne prime is a prime number of the form $2^p - 1$ (with p prime), such as $31 = 2^5 - 1$ or $131,071 = 2^{17} - 1$.

Mersenne primes are rare, and we don't even know if there are infinitely many of them. In January 2021 precisely 51 Mersenne primes were known, the largest being $2^{82,589,933} - 1$.

Expressed in binary, the two Mersenne primes above are

$$31 = 11111 \qquad 131,071 = 11111111111111111$$

with 5 and 17 repeated 1's, respectively. This makes it easy for a digital computer to do calculations using them. The Mersenne twister is based on a very large Mersenne prime, usually $2^{19,937} - 1$, and it replaces the numbers in the congruence by matrices over the field with two elements 0 and 1. It satisfies the statistical tests for substrings up to 623 bits long.

The GPS signal also incorporates a much lower-frequency signal that gives information about the satellite's orbit, its clock corrections, and other factors affecting the status of the system. This may sound very complicated, and it is, but modern electronics can handle highly complex instructions unerringly. There are good reasons for the complexity. It helps the receiver avoid accidentally locking onto some other random signal that happens to be floating around at the time, because it's highly unlikely that a stray signal would reproduce such a complex pattern. Each satellite is assigned its own personal pseudorandom code, so the same complexity ensures that the receiver doesn't confuse the signal from one satellite with that from another. As additional payoff, all of the satellites can transmit on the same frequency without jamming each other, which frees up more frequencies in our increasingly crowded radio spectrum. In military operations particularly, the enemy can't interfere with the system or send spurious signals. More generally, the US Department of Defense is in charge of the pseudorandom code, so it can control access to GPS.

*

As well as gradual drift of the atomic clocks, there are other sources of timing errors too, such as satellite orbits being slightly different in shape and size from the intended orbit. The ground station relays corrections to the satellite, which passes them on to users, ensuring that everything is in sync with the reference clocks at the US Naval Observatory. But the relativistic ones are where the mathematical action is greatest. So instead of old-fashioned Newtonian physics, we need Einstein's theories of relativity.[61]

In 1905 Einstein published a paper 'On the Electrodynamics of Moving Bodies'. He examined the relationship between Newtonian mechanics and Maxwell's equations for electromagnetism, finding these two theories to be incompatible. A central problem is that the speed with which electromagnetic waves propagate – the speed of light – is not only constant in a fixed frame of reference, but has the same constant value in a moving frame. If you shine a flashlight from a moving car, the photons travel at the same speed they would have done when the car was stationary.

In contrast, in Newtonian physics the car's speed would be added to the speed of light. Einstein therefore proposed modifying Newton's laws of motion to ensure that the speed of light is an absolute constant, which in particular implies that the equations for relative motion have to be modified. For this reason, the theory was named relativity, which is slightly misleading because the main point is that the speed of light is *not* relative. Einstein spent many years trying to incorporate gravity into his framework, eventually succeeding in 1915. These two related but distinct theories became known, respectively, as special and general relativity.

This isn't a textbook on relativity, so let me just skim through a few salient features, to give a very rough picture of what's involved. There isn't space to enter into the philosophical nuances, and even if there were, we'd be digressing, so please bear with me if I oversimplify.

In special relativity, the equations of motion are modified to ensure that the speed of light has the same value in any frame of reference that moves with constant velocity. This is achieved using Lorentz transformations, mathematical formulas named after the Dutch physicist Hendrik Lorentz, which describe how position and time change when different frames are compared. The main predictions are very strange from a Newtonian viewpoint. Nothing can travel faster than light; the length of an object shrinks as its speed increases, becoming arbitrarily small as the speed gets closer and closer to that of light; while this is happening, subjective time slows to a crawl; and mass increases without limit. Put crudely, at the speed of light, an object's length (in the direction of travel) shrinks to zero, time stops, and mass becomes infinite.

General relativity retains these ingredients, but also builds in gravity. However, instead of gravity being a force, as Newton modelled it, it's an effect of the curvature of spacetime, a four-dimensional mathematical construct combining three dimensions of space with one of time. Near any mass, such as a star, spacetime *bends*, making a kind of depression, but in four dimensions. A light ray or particle, passing nearby, deviates from a straight line as it follows the curvature. This creates the illusion of an attracting force between the star and the particle.

Both theories have been amply verified by highly sensitive experiments. Despite their rather bizarre features, they provide the best model of reality that physics has yet discovered. The mathematics of GPS must take account of relativistic effects, both from the velocity of the satellite and the Earth's gravity well, otherwise GPS would be useless. Indeed, the success of GPS, so corrected, is a sensitive test of the validity of both special and general relativity.

Most GPS users are either at fixed locations on the Earth's surface, or moving slowly – no more than the speed of a fast car, say. For this reason, the designers decided to broadcast information about the satellite orbits using a frame of reference rigidly attached to a rotating Earth, and assuming its rotation rate is constant. The shape of our planet, called a geoid, is approximately a slightly flattened ellipsoid of revolution.

When you're in your car, and the satellites are whirling overhead, they're obviously moving relative to you. Special relativity predicts that you will observe the satellite clock ticking more slowly than a reference clock on the ground. In fact, the satellite clock will fall behind by about 7 microseconds per day, thanks to relativistic time dilation. In addition to this, the apparent force of gravity at the height of the satellite orbits is less than on the ground. In terms of general relativity, spacetime up near the satellites is flatter – less curved – than it is near your car. This effect causes the satellite clocks to run *faster* than ground clocks. General relativity predicts that the satellite clocks gain on ground-based clocks by 45 microseconds per day. Putting these conflicting effects together, the clock on the satellite will tick faster than a ground clock by about $45 - 7 = 38$ microseconds per day. Such an error would become noticeable after two minutes, and your location would drift by about 10 km per day from the correct one. Within a day your satnav would place you in the wrong town, within a week in the wrong county, within a month in the wrong country.

Initially, the engineers and scientists working on the GPS project weren't sure that relativity actually mattered. Satellite speeds are fast by human standards, but a slow crawl compared to the speed of light. The Earth's gravity is tiny by cosmic proportions. But they did their best to estimate the sizes of these effects. In 1977, when the first prototype caesium atomic clock was placed in orbit, they still weren't sure how big these effects would be, or whether they should be positive or negative, and some didn't believe relativistic corrections would be needed at all. So the engineers incorporated into the clock a circuit that, at a given signal from the ground, could change its frequency to cancel out the predicted relativistic effects, if that turned out to be necessary. For the first three weeks they left this circuit turned off, and measured the clock frequency, observing it to be 442·5 parts per trillion higher, compared to a clock on the ground. The prediction from general relativity was an increase of 446·5 parts per trillion. Pretty much spot on.

*

There are many other uses of GPS, aside from the obvious one of locating positions (cars, commercial vehicles, hikers), and the military applications that led to the creation of GPS in the first place. I'll mention just a few.

You don't need to know where you are when you use an app to call for roadside assistance when your vehicle breaks down, because GPS does that for you. It's also used to prevent car theft, to perform mapping and surveying, to keep an eye on pets and elderly relatives, and to keep artworks safe. Major uses include ship and aircraft navigation and fleet tracking for transport companies. Now that most mobile phones have GPS receivers, they can tag your photographs with the location where you took them, tell you where a lost or stolen phone is, and call a cab. You can use GPS in conjunction with online mapping services like Google Maps, so the map automatically shows you where you are. Farmers can control driverless tractors, bankers can monitor financial transfers, travellers can track their luggage. Scientists can monitor the movements of endangered species and track environmental disasters such as oil spills.

How did we ever manage without GPS? It's astonishing how quickly a few bits of mathematical magic, enabling transformative (and highly expensive) technology, can change our lives.

12

De-Ising the Arctic

Greenland's ice sheet is melting much faster than
previously thought, threatening hundreds of millions
of people with inundation and bringing some of the
irreversible impacts of the climate emergency much
closer. Ice is being lost from Greenland seven times faster
than it was in the 1990s, and the scale and speed of ice
loss is much higher than was predicted.

The *Guardian*, December 2019

No, not icing: *Ising*. It's not a misprint. Just a bad pun.

The planet is heating up, it's dangerous, and it's our fault.
We know this because thousands of expert climate scientists,
running hundreds of mathematical models, have predicted it for
decades, and observations by equally competent meteorologists
confirm most of the important conclusions. I could spend the
rest of this book banging on about the purveyors of fake news
who try to convince us that there's nothing to worry about, and
contrast their inanities to the burgeoning evidence for the reality
of human-induced climate change, while explaining the many
points of fine detail that remain uncertain – but, as Arlo Guthrie
says about halfway through 'Alice's Restaurant', that's not what I
came to tell you about. Plenty of other people are doing that much
better than I can, and plenty of others are trying desperately to
stop them in case a few extremely rich people have to cease wreck-
ing the planet.

Climate change is inherently statistical, so any particular incident can be explained away as one of those weird things that just happen from time to time. If a coin is biased to toss heads three quarters of the time, any individual toss either gives heads or tails – just like an unbiased coin. So a single toss can't reveal the difference. Even a run of three or four consecutive heads can sometimes happen with a fair coin. However, if 100 tosses lead to 80 heads and 20 tails, it's pretty clear that it's not a fair coin.

Climate is similar. It's not the same as weather, which changes from hour to hour and day to day. Climate is a thirty-year moving average. Global climate averages that over the entire planet as well. It takes massive long-term changes on a planetary scale to alter climate. Yet high-quality worldwide temperature records go back about 170 years, and 17 of the 18 hottest years have occurred since 2000. That's no accident.

The statistical nature of climates make it easy for denialists to muddy the waters. Not being able to fast-forward the planet, climate scientists have had to rely on mathematical models to peer into the future, estimate how rapidly the climate is changing, work out what effects those changes might have, and examine what humanity can do about it if it gets its act together. The early models were fairly rudimentary, opening the door to objections from anyone who didn't like the predictions, although in retrospect it turns out that even these models got the rate of global temperature increase pretty much spot on, along with much else. Over the years they've been refined, and predicted temperatures now match reality in considerable detail over the last half-century. How much ice will melt as a consequence is less certain, and it appears to have been underestimated. The mechanisms involved aren't so well understood, and scientists have been under pressure for decades not to appear alarmist.

So far, I've concentrated on how mathematics, operating unsung behind the scenes, affects our daily lives. I've deliberately omitted a lot of important applications to science, especially theoretical science. But climate change does affect our daily lives – ask the Australians, who had to grapple with unprecedented bush fires early in 2020. Look at the record heatwaves all over the globe, the hundred-year floods now arriving every five or ten years. Look, oddly enough, at the occasional snap of extremely cold weather. It's somewhat counterintuitive that global warming can cause some places to get much *colder* than normal, but the explanation is simple. Global warming is about the average amount of heat energy going into the atmosphere, oceans, and land. Nobody said that everywhere would warm up uniformly.

As the planet's total heat energy goes up, fluctuations around the average get bigger, and those fluctuations can be colder than normal, as well as hotter. The point is that overall hotter wins. A sudden cold spell in one location isn't evidence that global warming is a hoax. Similarly, if your town is ten degrees colder than usual, but eleven towns elsewhere are one degree warmer, the average global temperature has gone *up*. If your town is ten degrees colder than usual today, but one degree warmer for eleven scattered days later on, the average global temperature has gone up, all else being unchanged. In fact, the average temperature in your town has also gone up.

The trouble is, we notice the sudden cold snap, but the compensating effects can be too small to impinge on our consciousness, too scattered, or just happening somewhere else. The highly unusual cold snaps that occurred in Europe and North America in recent years happened because the jet stream pushed cold air from the Arctic further south than normal. So the cold air that would normally have been circulating around the northern polar icecap ended up over the oceans, Greenland, and northern Canada and Russia. Why did all that cold air come south? Because the air in the polar regions was much *hotter* than normal, which displaced the cold air. Overall, the entire region affected got warmer – on average.

There's enough mathematics in climate modelling to fill a

book, but I don't want to tell you about that. Like Arlo, I'm just setting the scene for what I do want to tell you about.

*

All over the globe, ice is melting. In a few unusual places the amount of ice is going up, but everywhere else it's going down, fast. Glaciers are retreating, and the icecaps at both poles are shrinking. These effects threaten the water supplies of a couple of billion people, and the resultant rise in sea level will flood the homes of half a billion more, unless we stop it happening. So the physics and mathematics of melting ice has suddenly become of vital interest, on a personal level, to virtually everyone.

Scientists know a lot about melting ice. Along with water boiling and turning to steam, it's a classic example of a phase transition, a change in the state of matter. Water can exist in a variety of states. It can be a solid, a liquid, or a gas. Which state it's in depends mainly on temperature and pressure. At atmospheric pressure, water that's cold enough is a solid: ice. As it warms, and passes the melting point, it turns into a liquid: water. Heat it up a bit more, to its boiling point, and it becomes a gas: steam. Currently, 18 different phases of ice are known to science, the last, 'square ice', having been discovered in 2014. Three of these phases exist at normal pressures; the rest require much higher pressures.

Most of what's known about ice comes from laboratory experiments on relatively small quantities. What we urgently need to know about melting ice today is about extremely large quantities, in the natural environment. There are two intertwined ways to find out: observe and measure what's happening, and build theoretical models of the underlying physics. The key to real understanding is to bring both together.

One of the signs that polar ice, especially sea ice, is melting is the formation of melt ponds. The surface of the ice starts to

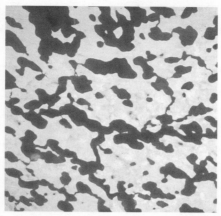

Dark melt ponds stand out against white Arctic ice.
Why do they make such intricate patterns?

melt, and small dark puddles besmirch the pristine whiteness of the ice, or often its less than pristine greyness from deposits of dust. The puddles are liquid water, and unlike ice they're dark, so they absorb sunlight instead of reflecting it away. Infrared radiation in particular warms the puddles faster than it would if they were still ice, so the puddles grow. When they get big enough, they merge to form larger puddles, large enough to count as ponds. These are melt ponds, and they make complex intricate shapes – blobs linked by thin strands, branching and spreading like patches of some weird fungus.

The physics of the growth of melt ponds is one vital feature of how sea ice behaves when it gets warmer. And that's exactly what's happening, especially to Arctic sea ice. What will happen to sea ice as the planet heats up is a vital part of the problem of understanding the impact of climate change. So it's natural for mathematicians to investigate models of melting ice, aiming to tease out some of its secrets. And so they do. Hardly a surprise. What is a surprise, though, is that one of the models now being studied isn't about melting ice at all. It's about magnetism, and it dates from 1920. Magnetic materials undergo their own kinds of phase transition, and in particular they lose their inherent magnetism if they get too hot.

This particular model has long been a poster child for phase transitions. It was invented by the German physicist Wilhelm Lenz, so of course everyone calls it the Ising model, because mathematicians and physicists always name things after the person most closely associated with it in their minds, who often isn't the actual inventor. Lenz had a student, Ernst Ising, and set him a problem for his PhD: solve the model and show that it has a magnetic phase transition. Ising solved it, and showed that it didn't. Nonetheless, his research kick-started an entire industry of mathematical physics, and greatly informed our understanding of magnets.

And, now, of melting ice.

*

Magnets are so familiar nowadays that we seldom wonder how they work. We use them to stick plastic pigs on the fridge door (well, in our house we do), to fasten covers on our mobile phones, and (using rather big ones) to detect the famous Higgs boson, which endows subatomic particles with mass. Everyday uses include computer hard drives and electric motors – the kind that automatically raise or lower your car window, or the kind that generate gigawatts of electrical power. Despite being ubiquitous, magnets are very mysterious. They attract or repel each other through some kind of invisible force field. The simplest and most familiar bar magnets have two poles, one near each end, called north and south. North and south poles attract each other, but two north poles repel each other, and the same goes for two south poles. If you try to push like poles of powerful little magnets together, you can feel them pushing back. If you try to pull unlike poles apart, you feel them trying to stick together. They affect each other even when they're not touching – 'action at a distance'.

With magnets, you can make objects levitate, even big ones like trains. Mysteriously, this field of force is invisible. You can't see a thing.

Humans have known about magnets for at least 2,500 years. They occur naturally in the mineral magnetite, an oxide of iron. A small lump of magnetite, known as a lodestone, can attract iron objects, and you can turn it into a compass by hanging it on a thread or floating it in water on a piece of wood. Lodestones were routinely used for navigation from about the twelfth century AD. Materials like this, that can be equipped with a permanent magnetic field, are said to be ferromagnetic. Most are alloys of iron, nickel, and/or cobalt. Some materials maintain their magnetism almost indefinitely, while others can be magnetised temporarily, but soon lose their magnetism again.

Scientists started to pay serious attention to magnets in 1820 when the Danish physicist Hans Christian Ørsted discovered a connection between magnetism and electricity. Namely, an electric current can create a magnetic field. William Sturgeon, a British scientist, made an electromagnet in 1824. The history of electromagnetism is too extensive to describe in any detail, but a key advance came from the experiments of Michael Faraday. These led James Clerk Maxwell to formulate mathematical equations for the electric and magnetic fields and the relation between them. The equations tell us, in a precise manner, that moving electricity creates magnetism, and moving magnetism creates electricity. Between them, they create electromagnetic waves that travel at the speed of light. Indeed, light is a wave of this kind. So are radio waves, X-rays, and microwaves.

One puzzling feature of ferromagnets is how they respond when heated. There's a critical temperature, called the Curie temperature. If you heat a ferromagnet above its Curie temperature, its magnetic field disappears. Not only that: the transition is abrupt. As the temperature approaches the Curie temperature, the magnetic field starts to drop dramatically, and decreases ever faster the nearer you get to the Curie temperature. Physicists call this type of behaviour a second-order phase transition. The big question is: why does it happen?

One important clue came with the discovery of the electron, a subatomic particle that carries a very tiny electric charge. An electrical current is a crowd of electrons in motion. Atoms have a nucleus, made of protons and neutrons, surrounded by a cloud of electrons. The number and arrangement of these electrons determine the chemical properties of the atom. Electrons also have a property called spin – it's a quantum property and they don't really rotate, but it has a lot in common with angular momentum, a fancy mathematical name for a feature of spinning bodies in classical physics. That tells us how powerful the spin is and in which direction it's happening – which axis the body is spinning about.

Physicists discovered experimentally that the electron's spin endows it with a magnetic field. Quantum mechanics being what it is – namely, weird – the electron's spin, measured about *any* specific axis, is always either 'up' or 'down'. These states correspond roughly to a tiny magnet with its north pole at the top and its south pole at the bottom or one the other way up. Before you measure the spin, it can be any combination of up and down simultaneously, which boils down to spinning about a different axis altogether, but when you observe the spin about your chosen axis, it always comes out as being up. Or down. One of those. That's the weird bit, and it's totally different from spin in classical physics.

The connection between the spin of an electron and its magnetic field goes a long way towards explaining not just why magnets lose their magnetism if they get too hot, but how they lose it. Before a ferromagnetic material is magnetised, the spins of its electrons are aligned randomly, so their tiny magnetic fields tend to cancel out. When the material is magnetised, either by an electromagnet or a close encounter with some other permanent magnet, the spins of its electrons are brought into alignment. They then reinforce each other, creating a detectable magnetic field on a large scale. If left undisturbed, this arrangement of electron spins persists, and you have a permanent magnet.

However, if you heat the material, the energy of the heat starts to jostle the electrons, flipping some of their spins. Magnetic fields pulling in different directions weaken each other, so the overall strength of the magnetic field drops. That explains the loss of magnetism in a qualitative way, but it doesn't explain why there's such a sharp phase transition, or why it always happens at a specific temperature.

Enter Lenz. He came up with a simple mathematical model: an array of electrons, each affecting its neighbours according to their relative spins. In the model, each electron is located at a fixed point in space, usually at a point in a regular lattice, like the squares of a large chessboard. Each model electron can exist in one of two states: $+1$ (spin up) or -1 (spin down). At any instant, the lattice is covered by a pattern of ± 1's. In the chessboard analogy, each square is either black (spin up) or white (spin down). Any pattern of black and white squares can occur, at least in principle, because quantum states are to some extent random, but some patterns are more likely than others.

PhD students are really useful for doing calculations or experiments that the supervisor prefers to avoid, so Lenz told Ising to solve the model. What the word 'solve' means here is subtle. It's not about the dynamics of how spins flip, or about individual patterns. What it means is: calculate the probability distribution of all possible patterns, and how this distribution depends on the temperature and any external magnetic field. A probability distribution is a mathematical gadget – often a formula – which in this case tells you how probable any given pattern is.

The supervisor has spoken, and if you want to get your PhD, you do what you're told. Or, at least, you give it your best shot, because supervisors sometimes set students problems that are too hard. After all, the reason for asking the student to solve the problem is that the supervisor *doesn't know the answer*, and often has no idea, beyond a vague gut feeling, of how hard it's going to be to find it.

So Ising buckled down to the task of solving Lenz's model.

*

There are some standard tricks, which PhD supervisors know about, and can suggest to their students. Really bright students discover these for themselves, along with ideas that never occurred to the supervisor. One of them is curious but generally true: if you want to work with a very large number, everything gets easier if you make it *infinite*. For example, if you want to understand the Ising model for a large but finite chessboard, representing a lump of ferromagnetic material of realistic size, it's mathematically more convenient to work with an infinitely large chessboard. The reason is that finite chessboards have edges, and these tend to complicate the sums because squares at the edge are different from those in the middle. This destroys the symmetry of the arrangement of electrons, and symmetry tends to make sums easier. An infinite chessboard has no edges.

The chessboard picture corresponds to what mathematicians and physicists call a two-dimensional lattice. The word 'lattice' means that the basic units, the chessboard's squares, are arranged in a very regular way – here in rows and columns, all perfectly aligned with their neighbours. Mathematical lattices can have any number of dimensions, while physical ones usually have dimension one, two, or three. The most relevant case for physics is the three-dimensional lattice: an infinite array of identical cubes, stacked together neatly like identical boxes in a warehouse. In this case the electrons fill a region of space, much like the atoms in a crystal with cubic symmetry, such as salt.

Mathematicians and mathematical physicists much prefer to

start with a simpler but less realistic model: a one-dimensional lattice, where the electron sites are arranged in a straight line at regular intervals, like integer points along the number line. Not very physical, but good to develop ideas in the simplest relevant setting. As the dimension of the lattice increases, the mathematical complications do the same. For example, there's one type of crystal lattice on a line, 17 in the plane, and a whopping 230 in three-dimensional space. So Lenz set his student the problem of finding out how models like this behave, and had the good sense to tell him to concentrate on the one-dimensional lattice. The student made enough progress for all such models to be called Ising models today.

Although the Ising model is about magnetism, its structure, and the way you think about it, belongs to thermodynamics. This area originated in classical physics, where it was about quantities like temperature and pressure in gases. Around 1905, when physicists finally became convinced that atoms exist and combine to make molecules, they realised that variables like temperature and pressure are statistical averages. They're 'macroscopic' quantities that we can easily measure, created by events occurring on a much smaller 'microscopic' scale. They're not actually visible in a microscope, by the way, even though today there are microscopes that can image individual atoms. These work only when the atom isn't moving. In a gas, a huge number of molecules are flying around, occasionally colliding and bouncing off each other. The bounces randomise their motion.

Heat is a form of energy, caused by the motion of the molecules: the faster they move, the hotter the gas gets, and up goes the temperature – which is different from heat: a measure of the quality of the heat, not the quantity. There are mathematical relationships between the positions and speeds of the molecules and the thermodynamic averages. This relationship is the subject of an area called statistical mechanics, which seeks to calculate macroscopic variables in term of microscopic ones, with special emphasis on phase transitions. For example, what is it about the behaviour of water molecules that changes when ice melts? And what does the temperature of the material have to do with it?

*

Ising's problem was similar, but instead of molecules of H_2O and ice turning into water as it got hotter, he was analysing electron spins and magnets losing their magnetism when they got hotter. Lenz had set up his model – the one we now call the Ising model – to be as simple as possible. As is common in mathematics, the model might be simple, but solving it isn't.

Recall that 'solving' the Ising model means calculating how statistical features of the array of tiny magnets vary with temperature. This boils down to finding the total energy of the system, and that depends on the pattern of magnetism – the number and arrangement of up and down spins, of black and white squares of the chessboard. Physical systems prefer to take up states with the lowest possible energy. This is why, for example, Newton's legendary apple fell: its gravitational potential energy got smaller as it dropped towards the ground. It was Newton's stroke of genius to realise that the same reasoning applies to the Moon, which is perpetually falling, but keeps missing the ground because it's also moving sideways. Doing the right sums, he showed that the same gravitational force explains both motions quantitatively.

Anyway, all the tiny magnets – the electrons with their spin directions – try to make their overall energy as small as possible. But how they do that, and what state they attain, depends on the temperature of the material. On a microscopic level, heat is a form of energy that causes molecules and electrons to jostle around randomly. The hotter the material gets, they more they jostle. In a magnet, the exact pattern of spins is constantly changing because of the random jostling, and that's why 'solving' the model leads to a statistical probability distribution, not a specific pattern of spins. However, the most likely patterns all look fairly similar, so we can ask what a typical pattern looks like at any given temperature.

The crucial part of the Ising model is a mathematical rule for how electrons interact, which specifies the energy of any pattern. The model makes a simplifying assumption that electrons interact only with their immediate neighbours. In a ferromagnetic interaction, that energy contribution is negative when the neighbouring electrons have the same spin. In antiferromagnetic systems it's positive when the neighbouring electrons have the same spin. There's also a further contribution to the energy caused by interaction of each electron with an external magnetic field. In simplified models, all the interaction strengths between neighbouring electrons have the same size, and the external magnetic field is set to zero.

The key to the mathematics is to understand how the energy of a given pattern changes when the colour of one square changes from black to white, or vice versa. That is, a single electron, in an arbitrary location, flips between +1 (black) and −1 (white). Some flips increase the total energy, others decrease it. Flips that decrease the total energy are more likely; however, flips that increase it aren't totally ruled out, because of the random thermal jostling. Intuitively, we expect the pattern of spins to converge to something with the lowest energy. In a ferromagnetic material, this should cause all the electrons to have the same spin, but in practice that's not quite what happens, because it would take too long. Instead, at moderate temperatures, there are distinct patches where the spins are almost perfectly aligned, creating a black-and-white crazy-paving pattern. At higher temperatures, random jostling beats the interactions between neighbouring spins, and the patches become so small that there's no relation between an electron's spin and the spins of its neighbours, so the pattern is chaotic and looks grey except for very fine black-and-white detail. At low temperatures the patches get bigger, leading to a more orderly pattern. These patterns never settle down completely;

there are always random changes. But, for a given temperature, the *statistical* features of the pattern do settle down.

What interests physicists most is the transition from separate patches of colour, an ordered state, to random grey chaos. This is a phase transition. Experiments on the ferromagnetic phase transition, from magnetised to demagnetised, show that below the Curie temperature the magnetic pattern is patchy. The sizes of the patches differ from each other, but they cluster around a specific typical size, or 'length scale', which gets smaller as the temperature increases. Above the Curie temperature, there are no patches: the two spin values are mixed up. What happens *at* the Curie temperature is what gets physicists excited. Now there are patches of various sizes, but there's no dominant length scale. The patches form a fractal, a pattern with detailed structure on all scales. A close-up of part of the pattern has the same statistical features as the entire pattern, so it's not possible to deduce the patch size from the pattern. There's no longer a well-defined length scale. However, the rate at which the pattern changes during the transition can be given a numerical measure, called the critical exponent. Experiments can measure the critical exponent very accurately, so it provides a sensitive test for theoretical models. A major goal of theorists is to derive models that give the correct critical exponent.

Computer simulations can't 'solve' the Ising model exactly – they can't provide a formula for the statistical features with a rigorous mathematical proof that it's correct. Modern computer algebra systems might help researchers to guess the formula, if there is one, but it would still need proof. More traditional computer simulations can provide solid evidence for or against the model matching reality. The holy grail for mathematical physicists (and physically inclined mathematicians, since the main problem is purely mathematical, though motivated by the physics) is to obtain *exact* results about statistical properties of the patterns of spins in the Ising model, especially how those properties change as the temperature passes through the Curie point. In particular, researchers seek a proof that a phase transition occurs in the model, and aim to characterise it through the critical exponent and the fractal features of the most probable patterns at the transition point.

*

Now the story gets more technical, but I'll try to give you the main ideas without worrying about the details. Suspend disbelief and go with the flow.

The most important mathematical gadget in thermodynamics is the 'partition function'. This is obtained by adding together, for *all* states of the system, a particular mathematical expression that depends on the state and the temperature. To be precise, we get this expression for any given state by taking the energy of that state, making it negative, and dividing by the temperature. Take the exponential of this, and add all such expressions together, for all possible states.[62] The physical idea here is that states with lower energies contribute more to this sum, so the partition function is dominated by – has a peak at – the most likely kind of state.

All of the usual thermodynamic variables can be deduced from the partition function by appropriate manipulations, so the best way to 'solve' a thermodynamic model is to compute the partition function. Ising found his solution by deriving a formula for the free energy[63] and deducing one for the magnetisation.[64] The formula looks impressive, but it must have come as a great disappointment to Ising, because after all those clever calculations, it tells us that when there's no external magnetic field, the material has no magnetic field of its own. Worse, this is true for any temperature whatsoever. So the model predicts no phase transition, and no spontaneous magnetisation of the supposedly ferromagnetic material.

It was immediately suspected that the main reason for this negative result is the simplicity of the model. Indeed, the finger of suspicion pointed to the dimensionality of the lattice. Basically, dimension one is too small to lead to realistic results. The next step, obviously, was to do the sums again for a two-dimensional lattice, but this was really hard. Ising's methods were inadequate. Only in 1944, after several breakthroughs that made such calculations more systematic and simpler, did Lars Onsager solve the two-dimensional Ising problem. This was a mathematical *tour de force*, with a complicated but explicit answer. Even then, he had to assume no external magnetic field.

The formula shows that now there *is* a phase transition, leading to a nonzero internal magnetic field below a critical temperature of $2k_B^{-1}J/\log(1 + \sqrt{2})$, where k_B is Boltzmann's constant from thermodynamics and J is the strength of the interactions between spins. For temperatures near the critical point, the specific heat goes to infinity, like the logarithm of the difference between the actual temperature and the true temperature, a characteristic of phase transitions. Later work derived various critical exponents as well.

*

What does all this fiddling about with electron spins and magnets have to do with melt ponds in Arctic sea ice? Melting ice is a phase transition, but ice isn't a magnet and melting isn't about spins getting flipped. How can there possibly be a useful connection?

If mathematics were wedded to the particular physical interpretation that gave rise to it, the answer would be 'there can't'. However, it's not. Not always, certainly. This is exactly where the mystery of the unreasonable effectiveness of mathematics comes into play, and why people who argue that inspiration from nature explains the *effectiveness* forget about the unreasonable bit.

Often the first clue to the possibility of this kind of portability, where a mathematical idea hops from one area of application to an apparently unrelated one, is an unexpected family resemblance

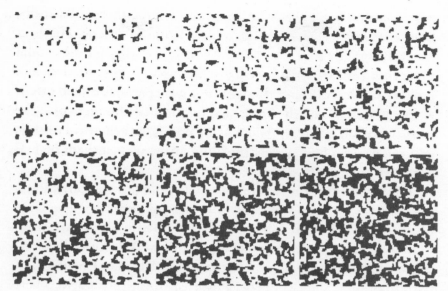

Simulation of melt-pond development based on the Ising model.

in a formula, a graph, a number, or a picture. Commonly this kind of resemblance turns out to be no more than a visual pun, an accident, a coincidence – full of sound and fury, signifying nothing. After all, there are only so many graphs or shapes to go round.

Just sometimes, however, it really is a clue to a deep relationship.

And that's how the research that I'm finally getting round to in this chapter got started. About ten years ago a mathematician named Kenneth Golden was looking at pictures of sea ice in the Arctic, and he noticed that they bore an uncanny similarity to pictures of patches of electron spins near the phase transition at the Curie point. He wondered whether the Ising model could be redeployed to shed light on how melt ponds form and spread. The model for ice is applied on a much larger scale, the up/down state of a tiny electron being replaced by the frozen/melted state of a region of the surface of the sea ice about one metre square.

It took a while for this thought to germinate into serious

mathematics, but when it did, it led Golden, working with atmospheric scientist Court Strong, to a new model for the effects of climate change on sea ice. He showed some Ising model simulations to a colleague who specialised in analysing images of melt ponds, and the colleague thought they were pictures of real ponds. A closer analysis of the statistical features of the images – such as the relation between the areas of ponds and their perimeters, which measures how wiggly the boundaries are – showed that the numbers match very closely.

The geometry of melt ponds is vital in climate research because it influences important processes on sea ice and the upper layers of the ocean. These include how the albedo of the ice – how much light and radiant heat it reflects – changes as it melts, how the floes break up, and how their sizes change. This in turn affects the pattern of light and dark beneath the ice, affecting photosynthesis in algae and the ecology of microbes.

Any acceptable model must agree with two main sets of observations. In 1998 the SHEBA expedition measured the sizes of melt ponds by taking images from helicopters. The observed probability distribution of pond sizes is a power law: the probability of finding a pond of area A is approximately proportional to A^k, where the constant k is about $-1\cdot5$ for ponds with areas between 10 and 100 square metres. This type of distribution is often indicative of fractal geometry. The same data, together with observations by the 2005 Healy–Oden Trans-Arctic Expedition HOTRAX, reveal a phase transition in the fractal geometry of melt ponds as they grow and coalesce, evolving from simple shapes into self-similar regions whose boundaries behave like space-filling curves. The fractal dimension of the boundary curves – the relation between area and perimeter – changes from 1 to about 2 at a critical pond area of about 100 square metres. This affects how the widths and depths of the ponds change, which in turn affects the size of the water–ice interface across which the ponds expand and, the bottom line, how fast they're melting.

The observed value of the exponent k is -1.58 ± 0.03, in good agreement with the SHEBA value of -1.5. The change in fractal dimension observed by HOTRAX can be calculated theoretically using a percolation model, and the larger dimension of roughly 2 turns out to be $91/48 = 1.896$ for this model. Numerical simulation of the Ising model leads to a fractal dimension very close to this.[65]

One interesting feature of this work is that the model operates on very fine length scales of a few metres. Most climate models have a length scale of several kilometres. So this kind of modelling is a radically new departure. It's still very much in its infancy, and the model needs to be developed into one that incorporates more of the physics of melting ice, absorption and radiation of sunlight, even winds. But it's already suggesting new ways to compare observations to mathematical models, and it starts to explain why melt ponds form such intricate fractal shapes. It's also the first mathematical model of the basic physics of melt ponds.

The *Guardian* report quoted in the epigraph to this chapter went on to paint a grim picture. The recent acceleration of ice loss in the Arctic, inferred from observations, not mathematical models, implies that the rise in sea level by 2100 will be two thirds of a metre, about two feet. This is seven centimetres (three inches) more than the Intergovernmental Panel on Climate Change (IPCC) previously forecast. About 400 million people will be at risk of flooding every year; 10% higher than the 360 million the IPCC previously predicted. Sea-level rise also makes storm surges more severe, causing further damage to coastal regions. In the 1990s Greenland was losing 33 billion tonnes of ice every year. Over the last ten years this has risen to 254 billion tonnes per year, and a total of 3.8 trillion tonnes of ice has been lost since 1992. About half of this loss is caused by glaciers moving faster and breaking up when they reach the ocean. The other half is caused by melting, mainly driven from the surface. So melt-pond physics is now of vital importance to everybody.

If the Ising metaphor can be made more precise, then all of the powerful ideas about the Ising model, won through strenuous efforts of generations of mathematical physicists, can be brought to bear on melt ponds. In particular, the connection to fractal geometry opens up fresh insights into the complex geometry of melt ponds. Above all, the story of Ising and the de-icing of the Arctic is a wonderful example of the unreasonable effectiveness of mathematics. Who could possibly have forecast, a century ago, that Lenz's model of the ferromagnetic phase transition could have anything to do with climate change and the ongoing disappearance of the polar icecaps?

13

Call the Topologist

Topological features are robust. The number of
components or holes is not something that should
change with a small error in measurement. This is vital
to applications.

Robert Ghrist, *Elementary Applied Topology*

Topology, a flexible type of geometry, was originally a very
abstract part of pure mathematics. Most people, among those
who have heard of it at all, still think it is, but that's starting
to change. That anything called 'applied topology' can exist
seems highly unlikely. It would be like teaching a pig to sing: what
would be remarkable is not that the animal sings well, but that it
sings at all. This assessment is right about pigs, but wrong about
topology. In the twenty-first century, applied topology is roaring
ahead, solving important problems in the real world. It's been
heading that way for some time, unnoticed, and has now reached
the point where it can safely be deemed a new branch of applied
mathematics. And it's not just a few applications of random bits
of topology: the applications are widespread and the topological tools involved cover large parts of the subject, including the
most sophisticated and abstract. Braids. Vietoris–Rips complexes.
Vector fields. Homology. Cohomology. Homotopy. Morse theory.
Lefschetz index. Bundles. Sheaves. Categories. Colimits.

There's a reason for this: *unity*. Topology itself has grown, in
just over a century, from a bunch of minor curiosities to a fully

integrated area of research and knowledge. It's now one of the major pillars on which the whole of mathematics rests. And where pure mathematics leads, applied mathematics usually follows. Eventually. (It happens the other way round, too.)

Topology studies how shapes deform under continuous transformations, and in particular, which features persist. Familiar examples of topological structures are the Möbius band, a surface with only one side, and knots. For about eighty years, mathematicians studied topology for its intrinsic interest, with no applications in mind. The subject became increasingly abstract, and esoteric algebraic structures known as homology and cohomology were invented to do things like counting the number of holes in a topological shape. It all seemed very obscure, with no practical implications.

Undaunted, mathematicians continued to work on topology, because of its central role in advanced mathematical thinking. As computers became more powerful, mathematicians began to look for ways to implement topological concepts electronically, allowing them to investigate very complex shapes. But they had to modify their approach to make it possible for computers to handle the calculations. The result, 'persistent homology', is a digital method for detecting holes.

At first sight, detecting holes seems far removed from the real world. But topology turns out to be ideal for solving some problems about networks of security sensors. Imagine a highly sensitive government facility, surrounded by woodland, attracting the attention of terrorists or thieves. To detect them approaching, you put motion sensors in the woods. What's the most efficient way to do this, and how can you be sure there are no holes in the coverage, through which the bad guys can pass unnoticed?

Holes? Of course! Call the topologist.

*

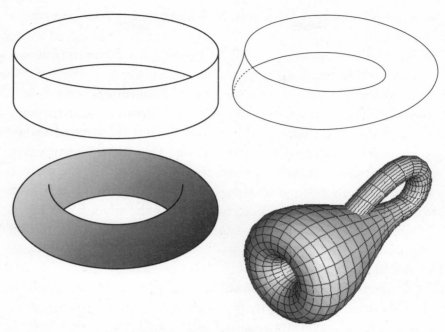

Top left: Cylinder. *Top right*: Möbius band.
Bottom left: Torus. *Bottom right*: Klein bottle.

When you first encounter topology, you're usually told about some basic shapes. They seem very simple, strange little toys. Some are whimsical, others downright weird. But it's whimsy with a point. As the great mathematician Hilbert once said: 'The art of doing mathematics consists in finding that special case that contains all the germs of generality.' Choose the right toy, and entire new subjects open up.

The first two toys in the picture can be made by taking a strip of paper and joining it end to end. The obvious way to do this gives a cylindrical band. A less obvious way is to give one end a 180° twist before joining them. This is a Möbius band, named after August Möbius who came across it in 1858, although it had previously been noticed by Gauss's student Johann Listing. It was Listing who first published the name 'topology', in 1847, but it was Gauss who presciently put him on to this nascent subject in the first place.

A cylinder has two separate edges, each a circle, and it has two distinct sides. You can paint the inside red and the outside blue, and the two colours never meet. In topology, what counts are properties of shapes that persist if you deform the shape continuously. You can stretch bits of it, compress them, twist things around, but not cut or tear, unless you join everything up again later. The uniform width of the cylindrical band in the picture is not a topological property: you can change the width by a continuous deformation. The circularity of the edges is not a topological property, for similar reasons. But being an edge, having two distinct edges, and having two distinct sides, are all topological properties.

Shapes that are considered the same when deformed have a special name: we call them topological spaces. The actual definition is highly abstract and technical, so I'll use more informal imagery. Everything I say can be made precise and given a decent proof.

We can use those topological properties to prove that a cylinder can't be deformed into a Möbius band. Even though they're both made by gluing the ends of a strip of paper together, they're different topological spaces. The reason is that the Möbius band has only one edge, and only one side. If you run your finger along the edge it goes round *twice* before getting back to the start, swapping from top to bottom because of the 180° twist. If you start to paint the surface red, you go all the way round, and then find you're painting the back of the part you've already painted, again because of the 180° twist. So the Möbius band has different topological properties compared to a cylinder.

The bottom-left shape is like a ring doughnut. Mathematicians call this shape a torus, by which they refer only to the surface, not the solid part where the dough would be. In this it's really more like an inflatable swimming aid. It has a hole. You can put your finger through it, or, in the case of a swimming aid, your entire body. But the hole isn't in the surface itself. If it were, the inflatable swimming aid would deflate and you'd sink. The hole manages to be in a place where the surface isn't. Fair enough: a broadband engineer sitting inside a manhole is also in a place where the surface isn't. But the manhole has edges, while the torus manages to have a hole even though it has no edges. Like the cylinder, it has two sides: the one we can see in the picture, and the one 'inside'.

The bottom-right shape is less familiar. It's called a Klein bottle, after the great German mathematician Felix Klein, and because it looks like a bottle. The name was probably a Germanic pun, because in German *Fläche* means 'surface' and *Flasche* is a 'bottle'. The picture is misleading in one respect: the surface seems to pass through itself. A mathematician's Klein bottle doesn't do that. The self-intersection arises because we naturally draw pictures as if the object is sitting in three-dimensional space. To get a Klein bottle that doesn't intersect itself, you either need to go to four dimensions, or better still, follow standard topological practice by discarding the need for a surrounding space completely. Then you can view a Klein bottle as a cylinder whose two circular ends have been joined together, but with one flipped over before making the join. To do this in three dimensions you have to poke that end inside and then open it out again, but you can also do it conceptually by just adding the rule that when you fall off one end you end up on the other one, reversing the direction round the circle. The Klein bottle has no edges, like a torus, but it's also like a Möbius band in having only one side.

We've now managed to distinguish all four of these topological spaces from each other. Either they have different numbers of edges, or different numbers of sides. Or different kinds of holes, if only we could say what we mean by a hole. This observation opens up one of the fundamental issues of topology. How can you tell whether two topological spaces are the same, or different? You can't just look at the shape, because that can be deformed. To a topologist, so the saying goes, a doughnut is the same as a coffee cup. You have to invoke *topological* properties that distinguish the spaces.

This can be hard.

*

The Klein bottle looks like the archetypal mathematician's toy. It's hard to see how it could ever be relevant to the real world. Of course, as Hilbert insisted, mathematical toys are useful not for themselves but for the theories they inspire, so the Klein bottle doesn't need to justify its existence directly. As it happens, however, this bizarre surface does show up in nature. It occurs in the visual system of primates – monkeys, apes, and of course us.

More than a century ago the neurologist John Hughlings Jackson discovered that the human cerebral cortex somehow contains a topographic map of the body's muscles. The cortex is the convoluted surface of the brain, so we all carry a map of our muscles inside our heads. This makes sense because the brain controls the contractions and relaxations of the muscles, causing us to move. A large part of the cortex is devoted to vision, and we now know that the visual cortex contains similar maps that operate the visual process.

Vision isn't just the eye acting like a camera and sending a photograph to the brain. It's far more complex, because the brain has to recognise the image as well as receive it. Like a camera, the eye has a lens to focus the incoming image, and the retina acts a bit like a film. In fact it's a lot closer to the way digital cameras record images. Light hits tiny receptors called rods and cones in the retina, and neural connections transmit the resulting signals to the cortex along the optic nerve – a bundle of numerous nerve fibres. These signals are processed along the way, but the cortex does the bulk of the analysis.

The visual cortex can be thought of as a series of layers, one on top of the next. Each layer has a specific role to play. The top layer, V1, detects boundaries between different parts of the image. This is the first step in *segmenting* the signal into its component parts. The boundary information is transmitted deeper into the cortex, being analysed at each step for the next kind of structural information, and it's then transformed for transmission to the next layer. Naturally, this description is a simplification, as are 'layers', and a lot of signals also travel in the reverse direction. The entire system creates, in our heads, a multicolour three-dimensional representation of the outside world – one so vivid and detailed that by default we assume it *is* the outside world. Which isn't entirely true, as a variety of visual illusions and ambiguities demonstrate. At any rate, eventually the cortex segments the image into parts that we can recognise as a cat, or Auntie Vera, or whatever. And then the brain can call up additional information, such as the name of the cat or Vera's recent lottery win.

The V1 layer detects boundaries using patches of nerve cells that are sensitive to edges that point in specific directions. The picture overleaf shows a portion of V1, obtained by optical recording from the visual cortex of a macaque monkey. Different shades of grey (colours in the original paper, so I'll call them that) correspond to neurons that fire when they receive data that indicates an edge in that orientation. The colours merge continuously from one shade to the next, except at certain isolated points where all colours exist nearby, in a kind of pinwheel configuration. These points are singularities of the orientation field.

This arrangement is constrained by topological properties of the orientation field. There are only two ways to arrange the series of colours round a singularity so that they change continuously: either the colours follow the sequence in a clockwise direction or they do so in an anticlockwise one. The picture shows examples of both. The presence of singularities is unavoidable because the cortex has to use many pinwheels to detect a complete line.

Colours (here shades of grey) show the orientation that creates the
most activity in each region of the cortex. The perceived orientation
changes smoothly except at singularities where all colours meet.

Now we ask how the brain combines this orientation information with information on how an edge is moving. Directions have
an arrow – north is opposite to south, although both lie on the
same straight line – and after a 180° rotation the arrow reverses.
You have to keep going through a full 360° before a direction gets
back to where it started. Edges don't have an arrow, and they
return to the same position after 180°. Somehow the cortex has
to make both these things work at the same time. If you draw a
loop round a singularity, the orientations vary continuously round
the loop, but the direction field must flip from a given direction
to the opposite one – say from north-pointing to south-pointing
– once or, more generally, an odd number of times. These statements are topological in nature, and they led Shigeru Tanaka to
conclude that the receptive fields connect to each other with the
topology of a Klein bottle.[66] This prediction has now been verified experimentally in various animals, among them owl monkeys,
cats, and ferrets, providing evidence that the organisation of the
visual cortex may be similar in many different mammals. The

experiments haven't been done with humans, for ethical reasons, but we're mammals, indeed primates. So it's plausible that, like macaques, we have Klein bottles in our heads, which help us perceive moving objects.

These ideas aren't just of interest to biologists. In the rapidly growing area of biomimetics, engineers take tips from nature to improve technology, leading to new materials and new machines. For instance, the curious structure of the lobster's eye played a crucial role in the invention of X-ray telescopes.[67] To focus a beam of X-rays you have to change their direction, but they're so energetic that a suitable mirror can deflect the beam only through a very small angle. Lobster evolution solved a similar problem for visible light millions of years ago, and the same geometry works for X-rays. New understanding of the V1 layer of the cortex in mammals can be transferred to computer vision, with potential applications to such things as self-driving vehicles and machine interpretation of satellite images for military or civilian purposes.

*

The central question in topology is 'what shape is this?' That is, 'which topological space am I looking at here?' It may sound a trite question, but mathematics presents us with topological spaces in innumerable ways – as pictures, as formulas, as solutions of equations – so it's not always easy to recognise what you've got. For example, it takes a topologist to see a Klein bottle in macaque V1. We made a stab at this problem when we observed that topological features distinguish the four spaces in my picture: cylinder, Möbius band, torus, Klein bottle. Towards the end of the nineteenth century, and early in the twentieth, mathematicians developed systematic ways to approach this question. The key idea is to define topological invariants: properties that you can calculate, which are the same for topologically equivalent spaces, but are different for at least some inequivalent spaces. Usually these aren't sensitive enough to distinguish all different spaces, but even a partial classification is useful. If two spaces have a different invariant of some kind, they definitely have different topologies. When considering the four shapes just referred to, invariants are things like 'how many edges?' and 'how many sides?'

Over the decades, some invariants turned out to be more useful than others, and some fundamentally important ones were constructed. The one I want to discuss now, in part because it has recently acquired some serious applications, is called *homology*. In essence, it counts how many holes of a given dimension a space has. In fact, it goes further than mere counting: it combines holes and non-holes together in a single algebraic object, called a homology group.

There's a very basic topological space that I've not yet mentioned: the sphere. Like the torus, when mathematicians use this word they mean the infinitely thin surface of a sphere, not a solid one. (That's a *ball*.) A sphere has no edges, like the torus and the Klein bottle. We can show that it's topologically different from both of these by looking at holes, or their absence.

Let's start with the torus. Visually, we can see that the torus has a whopping great hole through the middle. Spheres don't look remotely like that. But how do you define a hole mathematically, in a way that doesn't depend on a surrounding space? The answer is to look at closed curves on the surface. Every closed curve on a sphere forms the boundary of a region that, topologically speaking, is a disc – the interior of a circle.[68] Proving that is quite tricky, but it can be done, so let's assume it's true. On a torus, some closed curves also bound discs – but some don't. In fact, any closed curve passing 'through' the hole fails to bound a disc. Proving that is quite tricky too, but again let's go with the flow and assume it's OK. We've now shown that a sphere is topologically different from a torus, because 'closed curve' and 'bound a (topological) disc' are topological properties.

You can play this game in higher dimensions. For instance in

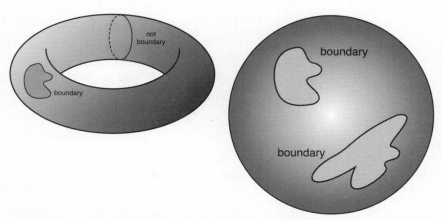

Left: On a torus, some closed curves are boundaries while others aren't.
Right: On a sphere, all closed curves are boundaries.

three dimensions you can replace 'closed curve' by '(topologically) spherical surface' and 'bound a disc' by 'bound a ball'. If you find a sphere that doesn't bound a ball, the space has some kind of three-dimensional hole. To go further and give some interpretation of what kind of hole, the early topologists discovered that you can add and subtract closed curves, or spheres. I'll discuss how it goes for curves on surfaces; higher dimensions are similar but messier.

Basically, you add two closed curves together by drawing them both on the same surface. To add a whole set of curves, draw them all. There are some technical refinements: it's often useful to draw an arrow round the curve to specify its orientation, and you can draw the same curve many times, or even a negative number of times. This is almost the same as drawing its reversal (same curve, opposite orientation) a positive number of times, in a sense I'll shortly explain.

A set of curves, labelled with numbers telling you how many times to draw them, is called a cycle. There are infinitely many possible cycles on a surface, but topologically, many of them are

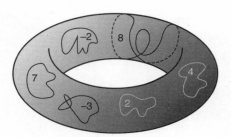

A cycle on a torus.

equivalent to many others. Now, I just said that minus a cycle is the same cycle with all arrows reversed. This isn't true as stated, because 'same' means 'identical', and they're not. But we can *make* them the same by playing a topological version of the trick that number theorists use in modular arithmetic. There, although 0 and 5 are different, we can pretend they're the same, for suitable purposes, and get the ring $\mathbb{Z}_5$ of integers modulo 5. In homology theory, we do the same kind of thing and pretend that any closed curve that bounds a disc is the same as the zero curve – draw *no* copies of it. Such a curve is called a boundary, and is said to be homologous to zero. The same idea extends to cycles: a cycle is homologous to zero if it's a combination of curves, each of which is a boundary.

We can add cycles C, D together to get $C + D$, as already described, and we can subtract them by reversing the arrow on D to get $C - D$, except that $C - C$ need not equal 0. This is annoying, but there's a way out: it's always *homologous* to zero. If we pretend that anything homologous to zero *is* zero, then we get a nice algebraic object called the homology group of the surface. In effect, we perform algebraic operations on cycles *modulo* (that is, ignoring) boundaries. Just as we do arithmetic (mod 5) by ignoring multiples of 5.

That's homology.

The homology group of a sphere is trivial: every cycle is

homologous to zero, and the group consists of 0 alone. The homology group of a torus isn't trivial: some cycles are not homologous to zero. It turns out that every cycle is homologous to an integer multiple of the one marked 'not boundary' in the picture, so the homology group of the torus is a disguised form of $\mathbb{Z}$, the integers. I won't do the sums and diagrams, but the homology group of the Klein bottle is $\mathbb{Z}_2 \times \mathbb{Z}_2$, pairs (m, n) of integers modulo 2. So it has some sort of hole, but it's a different kind of hole from the hole in (well, not in) a torus.

I ran you through the rather complicated construction of the homology group for a reason: to give you some idea of how topologists construct invariants. But the only message you need to take away is that every space has a homology group, it's a topological invariant, and you can use it to find out a lot about what shape the space is. Topologically speaking.

*

The homology group goes back to pioneering research by Enrico Betti and Poincaré at the end of the nineteenth century. Their approach was to *count* topological features, such as holes, but it was recast into group-theoretic language at the end of the 1920s by Leopold Vietoris, Walther Mayer, and Emmy Noether, and sweeping generalisations soon appeared. What I've called *the* homology group is just the first of an entire sequence of such groups, defining the algebraic structure of holes of dimension 1, 2, 3, and so on. There's also a dual notion of cohomology, and a related notion of homotopy, which is about how curves deform and join end to end, rather than how they relate to boundaries. Poincaré knew that this construction gives a group, which is usually not commutative. Algebraic topology is now a huge, highly technical subject, and new topological invariants continue to be discovered.

There's also a rapidly growing subject known as applied topology. As a new generation of mathematicians and scientists learned topology at their mothers' knee, they found it far less bizarre than the older generation had done. They spoke the language of topology fluently, and began to see new opportunities to apply it to practical problems. The Klein bottle in vision is an example from the frontiers of biology. In materials science and electronic engineering you find notions such as topological insulators: materials that can be switched from conducting electricity to insulating against it, by changing the topology of their electrical properties. Topological features, being preserved by deformations, are highly stable.

One of the most promising concepts in applied topology came into being when pure mathematicians were trying to write algorithms telling a computer how to calculate homology groups. They succeeded by rewriting the definition of a homology group in a manner more suited to computer calculations. These ideas then turned out to be a powerful new method for analysing 'big data'. This massively fashionable approach to all areas of science uses computers to seek hidden patterns in numerical data, and as the name suggests, these methods work best with very large quantities of data. Fortunately, today's sensors and electronics are amazingly good at measuring, storing, and manipulating gigantic amounts of data. Less fortunately, we often haven't a clue what to do with the data once we've collected it, but this is precisely where the mathematical challenges of big data lie.

Suppose you've measured millions of numbers, and you conceptually plot them as some sort of point cloud in a multidimensional space of variables. To extract meaningful patterns from the data cloud, you need to find the salient structural features. Paramount among these is the *shape* of the cloud. It's not feasible just to plot the points on a screen and eyeball them; you might be looking from the wrong angle, or important regions of points might be obscured by other points, or the number of variables might be too large for the visual system to process sensibly. Now, as we've seen, 'what shape is this thing?' is a fundamental question in topology. So it seems reasonable that topological methods might be useful; say to distinguish a roughly spherical data cloud from a toroidal one with a hole in it. We did a baby version of this for the FRACMAT project of Chapter 8. What mattered there was how compact the point cloud was, and whether it was round or cigar-shaped. Finer topological details weren't important.

You can't sort out topology by hand with a million data points: you have to use a computer. But computers aren't built to analyse topology. So the methods that the pure mathematicians had been developing for computer calculation of homology groups were redirected into the field of big data. And, as always, they didn't entirely do the job that was needed if you used them off-the-peg. They had to be modified to suit the new requirements of big data, the main one of these being that the shape of a data cloud isn't a well-defined thing. It depends, in particular, on the scale at which you observe it.

Imagine, for example, a hosepipe wound into a coil. Viewed from a moderate distance, a segment of the hosepipe looks like a curve, which topologically is a one-dimensional object. Closer up, it looks like a long cylindrical surface. Closer still, the surface acquires thickness; moreover, there's a hole that runs along the middle of the cylinder. Back off and view it from a distance but taking a wide-angle view, and the hose turns out to be coiled like a compressed spring. Coarsen your vision and the coil blurs into … a torus.

This kind of effect means that the shape of a data cloud isn't a fixed concept. So the homology group isn't such a great idea either. Instead, mathematicians asked how the *perceived* topology of the data cloud changes with the scale of observation.

Starting from a cloud and a selected length scale, you can create what topologists call a simplicial complex by joining two points by an edge whenever those points are closer together than the length scale. Then edges that are close together surround trian-gles, and triangles that are close together surround tetrahedrons,

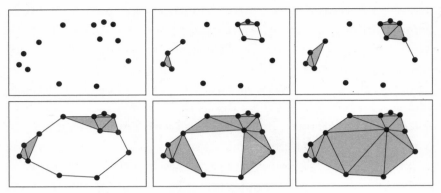

Joining data points separated by various distances creates
a sequence of triangulations, revealing holes of various
sizes. Persistent homology detects these effects.

and so on. A multidimensional tetrahedron is called a simplex, and a collection of simplexes, joined together in some manner, is a simplicial complex. A simpler name, 'triangulation', will do here. Just remember that the triangles can be of any dimension.

When you have a triangulation, there are mathematical rules to compute the homology. But now, the triangulation depends on the scale of observation. So the homology depends on that as well. The interesting question about shape then becomes: how does the homology of the triangulation change as the scale changes? The most important features of the shape ought to be less susceptible to change than the more transient features that depend sensitively on the scale. So we can focus on those aspects of the homology group that *persist* when the scale changes. The resulting gadget, not just a homology group but a family of them, one for each scale, is known as persistent homology.

Here the sequence of six pictures shows which points join up at various scales. As the length scale increases, and we look at coarser structures, an initial cloud of isolated points starts to form small clumps, one of which has a small hole. That hole fills in and the clumps grow. Then the clumps join into a ring to reveal

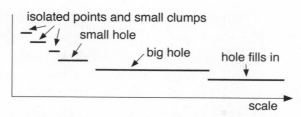

Barcode for persistent homology shows which structures
persist over which scales. (Schematic.)

a big hole. This starts to thicken up, but remains a big hole, until
the scale gets so large that everything gets filled in. This picture is
schematic and details that a computer algorithm would add have
been suppressed for clarity. The dominant feature, occurring for
the biggest range of length scales, is the big hole in the middle.

Notice that this description includes distance information as
well as topology. Technically a topological transformation need
not preserve distances, but in data analysis the actual values of
the data are important, as well as the overall topological shape.
For this reason, persistent homology pays attention to metric
properties as well as topological ones. One way to present the
information provided by persistent homology is to construct a
barcode which uses horizontal lines to represent the range of
scales over which particular homological features (such as holes)
persist. For example, the barcode for the point cloud in the picture
might look a bit like the one in the above picture. The barcode is a
schematic summary of how topology varies with scale.

*

Persistent homology and its barcodes are all very elegant, but
what are they good for?

Imagine you're running a business, and your offices are situated
in a clearing in a wood. Burglars can approach unseen through

the wood. So you set up an array of sensors, each of which can detect motion and communicate with neighbouring sensors, and switch this on at night. If anyone approaches, authorised or not, the sensors will trigger an alarm and your security people can go and investigate. Or imagine you're a General running a military base in an area where terrorist groups are active. You'd do something similar, carrying weapons.

How can you ensure that the sensor coverage is adequate, without gaps where a criminal or terrorist might sneak through?

If you're using a small number of sensors, you can map their distribution and eyeball the result. With larger numbers, or varying limitations caused by the terrain, this becomes less practical. So you need a way to detect holes in the sensor coverage … Detect *holes*? This sounds like a job for persistent homology. And, indeed, this is one of the many uses now being made of this new idea. A similar application is 'barrier coverage': decide whether a collection of sensors completely surrounds an important building or complex. 'Sweeping coverage' deals with sensors that can move; a domestic or commercial version applies to robot vacuum cleaners. Will they clean the entire floor?

A more scientific application is carried out in conjunction with the sliding-window method for reconstructing dynamical attractors that I mentioned in Chapter 8. Persistent homology can detect when the topology of the attractor changes significantly. In dynamical systems theory this effect is called bifurcation, which signals a major change in the dynamics. One important application is to find out how the Earth's climate has changed over millions of years from warm periods to ice ages, even to the total ice coverage of Snowball Earth. Jesse Berwald and colleagues have shown that the barcodes of sliding-window data clouds do an excellent job of identifying changes in the overall climate regime.[69] Other cases where the same method applies to different physical systems are vibrations in machine tools used in manufacturing, which leave unwanted flaws and marks on the surfaces of the objects being made. Firas Khasawneh and Elizabeth Munch have shown that time series measurements of the cutting tool can pick up this kind of vibration, known in the trade as 'chatter'.[70] There are also applications to medical imaging, such as the detection of biphonation in laryngeal video endoscopy, by Christopher Tralie and Jose Perea.[71] This effect occurs when a vocal cord produces two frequencies of sound at the same time, and it can indicate lesions or paralysis. Endoscopy is the insertion of a camera on the end of a fibre optic cable, up the nose and down the throat. Saba Emrani and others[72] have employed barcodes on audio data to detect wheezing in patients, an abnormal high-pitched sound that can be indicative of partially blocked airways or lung diseases such as asthma, lung cancer, and congestive heart failure.

Got a problem with your data? Need help urgently?

Call the topologist.

14

The Fox and the Hedgehog

The fox knows many things, but the hedgehog knows one big thing.

Attributed to Archilochus, c. 650 BC.

When seeking inspiration for this book, I stumbled across the phrase: *Πόλλ' οἶδ' ἀλώπηξ, ἀλλ' ἐχῖνος ἓν μέγα*. I did Latin at school, not Greek, but mathematicians know their Greek letters. Even I can recognise 'echinos' and 'mega', and make an educated guess, so it's something about a big hedgehog. It actually translates as: 'The fox knows many things, but the hedgehog knows one big thing.' It was probably coined by the ancient Greek poet Archilochus, but that's uncertain.

Should I be a fox or a hedgehog? Should I try to describe a selection of the innumerable amazing developments in mathematics over the past fifty years, and how they're used? Or should I focus on *one big thing*?

I decided to do both.

You've waded through the foxy bit, all thirteen chapters of it. Now comes the hedgehoggy bit, to sum it all up.

Looking back at the topics that we've covered, I'm amazed at the richness and variety of the different areas of mathematics that have now burrowed their way into the systems and devices that characterise life in the early twenty-first century. Not just for the well-heeled in a Western democracy, though they probably benefit more than less wealthy people, but for billions of people in every country in the world. The mobile phone has brought modern communications to developing countries. It's now ubiquitous, and it's changed everything. Not always for the better, but change is double-edged. Without mathematics, and a lot of people trained to use it at an advanced level, there would be no mobile phones.

I'm also aware of the huge number of applications that I've not had space to mention. The ones you've been reading aren't necessarily the best, the most important, the most impressive, or the most valuable. They're just a collection of applications that appealed to me because they take good, usually new, mathematics and use it in an area that comes as a surprise, because that kind of mathematics wasn't invented for that purpose at all. I also aimed at variety – I don't think it would have made sense to spend 90% of the book on applied partial differential equations, say, although it would have been easy to find enough material and to justify their importance. I wanted to show you the diversity and the broad scope of today's uses of my subject, in addition to establishing its relevance to humanity at large.

To salve my conscience, I'll briefly mention a few of the hundreds of other applications that I could have told you about instead. Even these are just the tip of the iceberg. While researching this book I collected a file, and these examples are taken from that. They're in no special order.

Forecasting flood levels.

Large data analysis and Lyme disease.

How many shakes does it take to get ketchup out of the bottle?

How to optimise the use of wood in sawmills.

How best to insulate a house or a pipe.

Detecting bias (racial, gender) in algorithms.

Rigidity of engineering frameworks, such as steel frames for buildings.

Computer recognition of cancer cells.
Improving consistent thickness in glass sheet manufacture.
Production of carbon dioxide when concrete cures.
Designing master key systems for office buildings.
Computer modelling of a virtual heart.
Designing buildings to resist hurricanes.
Finding ancestral relations between species.
Planning the movements of industrial robots.
Epidemiology of diseases in cattle.
Traffic queues.
Constructing a weather-smart electricity grid.
Improving the resistance of communities to hurricane storm
    surges.
Underwater communications cables.
Detecting landmines in countries where wars have ended.
Predicting how dust from volcanoes will move, to assist
    airlines.
Reducing voltage fluctuations in power grids.
Improving the efficiency of tests for viruses during the
    COVID-19 pandemic.

Each of these topics could reasonably have merited its own chapter. They add further examples of the sheer diversity of ways in which mathematics is used for the benefit of everyone on this planet.

*

As these examples, and the others that I've discussed in more detail, make clear, the sheer variety of applications of mathematics is mind-boggling – especially when you realise that much of it was originally created either with a different objective in mind, or just because some mathematician somewhere at some time thought it might be interesting to take a look. This once more raises the deep philosophical issue that puzzled Wigner back in 1959. And it remains – to me, at least – just as puzzling now as it was then. If anything, more so. Wigner was mainly focused on the unreasonable effectiveness of mathematics in theoretical physics, but now we find that it's unreasonably effective in a far wider, and more immediate, range of human activities. Most of those have even less apparent connection with anything mathematical.

Like Wigner, I'm not convinced by the explanation that many people suggest: mathematics derives from the real world, and must therefore be effective in the real world. As I've said already, I think this misses the point, although it does a good job of explaining *reasonable* effectiveness. The stories I've told in *What's the Use?* illustrate some of the features that make mathematics useful in areas apparently unrelated to its origins. The mathematician and philosopher Benjamin Peirce defined mathematics as 'the science which draws necessary conclusions'. Given *this* and *that*, what's going to happen? This is a very general issue, common to most problems that arise in the outside world. Because mathematics nowadays is very general, it provides a kit of useful tools for answering such questions, just waiting to be brought to bear. You don't need to envisage every possible use for a hammer in order to decide that a hammer might be worth having. Being able to bang stuff together, or knock it apart, is a general technique that's likely to be widely applicable. A hammer works on one task, so it may well work on others. A mathematical method that has been perfected on one application can often be transferred, suitably modified, to others.

Another definition of mathematics that I like is Lynn Arthur Steen's 'the science of significant form'. Mathematics is about *structure*. It's about how to leverage structure to understand a problem. Again, this point of view is one of considerable generality, and experience shows that it can cut to the heart of the matter.

A third definition, proposed in desperation, is that mathematics is 'what mathematicians do'. To which we add that a mathematician is 'someone who does mathematics'. I think we can do better than a tautology. Is business 'what business people do' and a business person 'someone who does business'? Yes, but there's more to it than that. What makes someone a successful entrepreneur isn't doing business as such: it's spotting an *opportunity* to do business that others have overlooked. Similarly, a mathematician is someone who spots an opportunity to do mathematics that others have overlooked.

The way to do that is to think mathematically.

Over the centuries mathematicians have evolved reflex ways of thinking that strike at the heart of problems. What's the natural context for the problem? What's the space of possibilities? What's the natural structure in which to express the relevant properties? Which features are essential, and which are irrelevant fine detail or distractions that can be thrown out? How do you throw them out? What's the natural structure on what remains? The mathematical community has honed those methods on countless hard problems, refined them into elegant and powerful theories, and challenged them with problems from the real world. They've become increasingly general, interconnected, powerful, and portable.

Maybe the effectiveness of mathematics isn't so unreasonable.

Maybe it's not a puzzle at all.

*

Imagine a world without mathematics.

I hear a lot of people cheering wildly, and they have my sympathy, because there's no reason why something that appeals to me should also appeal to you. But I'm not talking about you personally avoiding having to learn the subject. It's not just about you.

Suppose that out there in the wide, wide universe is an alien civilisation that consumes enormous amounts of mathematics. I mean this literally. Some physicists argue that mathematics is unreasonably effective in explaining the universe because the universe is *made from* mathematics. Mathematics isn't a human technique for understanding things: it's real, some ethereal substance that's built into everything there is.

Personally, I think this view is crazy, and it trivialises the philosophical conundrum, but the aliens know that I'm wrong. They discovered a billion years ago that the universe really is made of mathematics. And their civilisation consumes it in vast quantities, just as we're using up many of the Earth's resources. In fact, the aliens have consumed so much mathematics that they would have run out long ago, were it not for a simple solution. Their technology is extremely advanced and their attitudes are extremely aggressive, so they send fleets of gigantic spaceships among the stars, armed to the teeth, seeking out new life and helping themselves to its mathematics.

The mathivores are coming.

When they arrive at a new world, they *eat* all of its mathematics. Not just the ideas, but the ethereal substance itself – and everything that has ever depended on the subject disappears too, deprived of its support. Mathivores prefer the more refined foodstuffs, so they start with really advanced mathematics and eat their way down to the more prosaic stuff. They generally leave when they get to long multiplication, because they don't find basic arithmetic terribly palatable, so civilisation on the world they've attacked doesn't break down completely. It is, however, a pale shadow of its former glory, and the Galaxy is littered with planets whose indigenes have been propelled back into the Dark Ages with no prospect of escape.

If the mathivores arrived tomorrow, what would *we* lose?

We probably wouldn't notice when pure mathematics at the research frontiers disappeared. Although some of that would probably become vital in a century's time, it's not essential now. But as the mathivores work their way down from the ivory towers, important things start to disappear. The first to go are computers, mobile phones, and the Internet, the mathematically most sophisticated products on the planet. Anything involving spaceflight goes next: weather satellites, environmental satellites, communications satellites, satnav, aircraft navigation, satellite TV, solar flare observatories. Electrical power stations cease to function. Industrial robots grind to a halt, manufacturing industries die, and we go back to brooms instead of vacuum cleaners. No jet aircraft: we can't design them any more without computers, and we need aerodynamics to figure out how to make them stay aloft. Radio and TV vanish in puffs of alien smoke, because those technologies rely on Maxwell's equations for electromagnetic radiation – radio waves. All large buildings fall down, because their design and construction depends heavily on computer methods and elasticity theory to ensure structural integrity. No skyscrapers, no big hospitals, no sports stadiums.

History is running backwards. Already we've reverted to what life was like a century ago, and the mathivores have hardly started.

Some losses are arguably good: nuclear weapons, for instance, and most other military applications of mathematics, although we also lose the ability to defend ourselves. The subject itself is neutral: what's good or bad depends on what people do with it.

Some losses are equivocal: banks shut down all investment in the stock market because they lose the ability to forecast what it will do, thereby minimising their financial risk. Bankers don't like risks, except those they're unaware of until the financial system falls to bits. This reduces our self-destructive obsession with money, but it also stops a lot of useful projects getting finance.

Most losses are bad. Weather forecasting reverts to licking your finger and holding it up to tell which way the wind's blowing. Medicine loses its scanners and its ability to model the spread of epidemics, though it keeps anaesthetics and X-rays. Anything that depends on statistics is as dead as the dodo. Doctors can no longer assess the safety and effectiveness of new medicines and treatments. Agriculture loses the ability to assess new breeds of plants and animals. Manufacturing industry can no longer perform effective quality control, so everything you buy – from the limited range of goods still available – is unreliable. Governments lose the ability to predict future trends and demands. They might not have been terribly good at it anyway, but now they get much worse. Our communications revert to the primitive, not even the telegraph. Sending letters on horseback is the fastest we can manage.

By this point the current human population has become impossible to support. None of the clever tricks we've been using to grow more food and to transport goods across the oceans work any more. We have to revert to sailing ships. Disease runs rampant as billions starve to death. The End Times are here, and Armageddon awaits as the few survivors do battle for what little is left of our world.

*

You may feel my scenario is an exaggeration. I'd argue passionately that the only thing I've exaggerated is the metaphor of mathematics as an edible substance. We really do rely on mathematics for almost everything that keeps our planet ticking. The daily lives of people who think mathematics is useless unknowingly rely on the activities of those who know that's not true. It's not remotely their fault: those activities take place behind the scenes, where no one except a specialist is ever likely to become aware of them.

I'm not saying 'without mathematics we'd still be in caves', because I'm sure that without mathematics we'd have found other ways to advance. I'm absolutely not claiming that mathematics *alone* should be given the credit for the advances that we've made. Mathematics is at its most useful in conjunction with everything else that humanity can bring to bear on the problems that it faces and the objectives it envisages. But we are where we are because mathematics, along with all those other things, has put us here.

Now we've embedded mathematics so deeply in our technological and social structures that we'd be in a terrible state without it.

In the opening chapter I cited six features of mathematics: reality, beauty, generality, portability, unity, and diversity. Together, I claimed, these lead to utility. How do those remarks stack up now that you've read Chapters 1–13?

Many of the mathematical ideas we've covered originated in the real world. Numbers, differential equations, the TSP, graph theory, the Fourier transform, the Ising model. Mathematics takes inspiration from nature, and is all the better for it.

Other strands of the subject arose largely because of the pure mathematician's sense of beauty. Complex numbers, invented because it's ugly when some numbers have two square roots and others have none. Modular arithmetic, elliptic curves, and other parts of number theory, because people enjoyed looking for numerical patterns. The Radon transform, an interesting question in geometry. Topology, having little contact with reality for a century, but central to the mathematical edifice because it's about continuity, which is fundamental.

The urge to generalise is visible everywhere. Euler didn't just solve the puzzle of the Königsberg bridges; he solved every puzzle of the same kind and created a new area of mathematics: graph theory. Codes based on modular arithmetic led to questions of computational complexity and whether P = NP. Complex numbers inspired Hamilton's quaternions. Analysis was generalised to functional analysis, replacing finite-dimensional spaces by infinite-dimensional function spaces and functions by functionals and operators. Mathematicians invented the Hilbert spaces of quantum theory long before physicists found a use for them. Topology started with toys like Möbius bands and exploded into one of the deepest and most abstract areas of human thought. Now it's starting to pay off in everyday life, too.

Many of the methods we've encountered are portable, so they get used all over the place, no matter where they originated. Graph theory turns up in medical problems about kidney transplants, in the TSP, in quantum codes (expander graphs) that can protect our data against attacks by a quantum computer, in satnav's ability to choose a sensible route. The Fourier transform was originally devised to study the flow of heat, but its cousins include the Radon transform, used in medical scanners, the discrete cosine transform in JPEG image compression, and the wavelets that the FBI uses to store fingerprints efficiently.

The unity of mathematics is also a strand that runs throughout my stories. Graph theory segues into topology. Complex numbers appear in number theoretic problems. Modular arithmetic inspires the construction of homology groups. Satnav unites at least five distinct branches of mathematics in one application, from pseudorandom numbers to relativity. Dynamics helps put satellites in orbit and suggests a new quality control method for spring wire.

Diversity? Between them, the chapters in this book feature dozens of different areas of mathematics, usually in combination. They range from the numerical to the geometric, from irrational numbers to Klein bottles, from fair cake division to climate models. Probability (Markov chains), graphs, and operations research (Monte Carlo methods) join forces to increase patients' chances of securing a kidney transplant.

As for utility: the range of applications is if anything even more diverse, from movie animation to medicine, from spring manufacture to photography, from Internet commerce to airline routing, from mobile phones to security sensors. Mathematics is everywhere. And I've shown you only a tiny part of what's out there, running the world, unseen and unheralded. I have no idea what most of it is. Many of the best ideas are commercial secrets, anyway.

When it comes to the crunch, that's why we need as many of us as possible to have as extensive a grasp of mathematics as possible. Not just for our own personal benefit; I accept that for most of us, much of what we're taught about mathematics isn't directly useful. But that's true of everything. I studied history at school, and it certainly gave me a better feel for the culture in which I live, as well as feeding me with colonialist propaganda that now seems increasingly biased. But I don't *use* history in my work or my life. I find it interesting (more so, as I get older), I'm glad there are historians who do use it, and I wouldn't dream of recommending that it shouldn't be taught. But the evidence is clear: mathematics is *essential* to today's way of life. Moreover, it's very hard to predict what bits might be useful to us tomorrow. My bathroom tiler Spencer didn't think π was much use, until he needed it.

Mathematics, properly understood as the rich and creative subject that it really is, and not the low-level caricature that so many imagine, is one of the greatest achievements of humanity. Not just intellectually, but practically. Yet we hide it in the dark. It's time to bring it out into the light, before real-world analogues of my sci-fi mathivores try to take it away from us.

Yes, the fox knows many things, but mathematicians know One Big Thing. It's called mathematics, and it is remaking our world.

Notes

1 In 2012 the accountancy company Deloitte carried out a survey: *Measuring the Economic Benefits of Mathematical Science Research in the UK*. At that time, 2·8 million people were employed in mathematical science occupations: pure and applied mathematics, statistics, and computer science. The mathematical sciences contributed £208 billion (gross value added) to the UK economy in that year – just under £250 billion in 2020 money, around $300 billion. Those 2·8 million people made up 10% of the British workforce, and contributed 16% of the economy. The largest sectors were banking, industrial research and development, computer services, aerospace, pharmaceuticals, architecture, and construction. The report's examples include smartphones, weather forecasting, healthcare, movie special effects, improving athletic performance, national security, managing epidemics, Internet data security, and making manufacturing processes more efficient.

2 https://www.maths.ed.ac.uk/~v1ranick/papers/wigner.pdf

3 The formula is

$$\frac{1}{\sigma\sqrt{2\pi}}e^{-\frac{1}{2}\left(\frac{x-\mu}{\sigma}\right)^2}$$

where x is the value of the random variable, μ is the mean, and σ is the standard deviation.

4 Vito Volterra was a mathematician and physicist. In 1926 his daughter was courting Umberto D'Ancona, a marine biologist, and later they married. D'Ancona had discovered that during the First World War, the proportion of predatory fish (sharks, rays,

swordfish) that fishermen were catching increased, even though they were doing less fishing overall. Volterra wrote down a simple calculus-based model for how the populations of predators and prey change over time, which showed that the system goes round and round in a cycle of predator explosions and prey crashes. Crucially, *on average* the number of predators increases, proportionately, more than the number of prey.

5   No doubt Newton used physical intuition as well, and historians tell us that he probably pinched the idea from Robert Hooke, but there's no point in being a one-trick pony.

6   www.theguardian.com/commentisfree/2014/oct/09/virginia-gerrymandering-voting-rights-act-black-voters

7   Time wasn't the only issue. At the Constitutional Convention of 1787, which led to the Electoral College system, though not by that name, James Wilson, James Madison, and others felt that a popular vote would be best. However, there were practical problems about who would be allowed to vote, with big differences of opinion between Northern and Southern states.

8   In 1927 E.P. Cox used the same quantity in palaeontology to assess how round sand grains are, which helps distinguish windblown sand from waterborne sand, providing evidence for environmental conditions in prehistoric times. See E.P. Cox. 'A method of assigning numerical and percentage values to the degree of roundness of sand grains', *Journal of Paleontology* 1 (1927) 179–183. In 1966 Joseph Schwartzberg proposed using the ratio of the perimeter of a district to the circumference of the circle of the same area. This is the reciprocal of the square root of the Polsby–Popper score, so it ranks districts in the same way, though with different numbers. See J.E. Schwartzberg. 'Reapportionment, gerrymanders, and the notion of "compactness"', *Minnesota Law Review* 50 (1966) 443–452.

9   By enclosing a hill, a curved surface, she crammed even more area into her circle.

10 V. Blåsjö. 'The isoperimetric problem', *American Mathematical Monthly* 112 (2005) 526–566.

11 For a circle of radius r,

circumference (= perimeter) = $2\pi r$

area = πr^2

perimeter^2 = $(2\pi r)^2 = 4\pi^2 r^2 = 4\pi(\pi r^2) = 4\pi \times$ area

12 N. Stephanopoulos and E. McGhee. 'Partisan gerrymandering and the efficiency gap', *University of Chicago Law Review* 82 (2015) 831–900.

13 M. Bernstein and M. Duchin. 'A formula goes to court: Partisan gerrymandering and the efficiency gap', *Notices of the American Mathematical Society* 64 (2017) 1020–1024.

14 J.T. Barton. 'Improving the efficiency gap', *Math Horizons* 26.1 (2018) 18–21.

15 In the early 1960s John Selfridge and John Horton Conway independently found an envy-free method of cake division for three players:

1. Alice cuts the cake into three pieces that she considers of equal value.

2. Bob either passes, if he thinks two or more pieces are tied for largest, or trims what he considers to be the largest piece to create such a tie. Trimmings are called 'leftovers' and set aside.

3. Charlie, Bob, and Alice, in that order, choose a piece that they think is largest or tied largest. If Bob didn't pass in step 2 he must choose the trimmed piece, unless Charlie chose it first.

4. If Bob passed at step 2 there are no leftovers and we're done. If not, either Bob or Charlie took the trimmed piece. Call this person the 'non-cutter' and the other the 'cutter'. The cutter divides the leftovers into three pieces that he considers equal.

5. Players choose one of these pieces in the order non-cutter, Alice, cutter. No player has any reason to envy what another player receives: if they do, they got their tactics wrong and should have chosen differently. For a proof, see: en.wikipedia. org/wiki/Selfridge-Conway_procedure

16 S.J. Brams and A.D. Taylor. *The Win-Win Solution: Guaranteeing Fair Shares to Everybody*, Norton, New York (1999).

17 Z. Landau, O. Reid, and I. Yershov. 'A fair division solution to the problem of redistricting', *Social Choice and Welfare* 32 (2009) 479–492.

18 B. Alexeev and D.G. Mixon. 'An impossibility theorem for gerrymandering', *American Mathematical Monthly* 125 (2018) 878–884.

19 B. Gibson, M. Wilkinson, and D. Kelly. 'Let the pigeon drive the bus: pigeons can plan future routes in a room', *Animal Cognition* 15 (2012) 379–391.

20 My favourite example is a politician who made a huge fuss about money being wasted on what he called 'lie theory' – pronouncing 'lie' as in 'untruth', which is what he thought it was about. Not so. Sophus Lie (pronounced 'lee') was a Norwegian mathematician, whose work on continuous groups of symmetries (Lie groups) and associated algebras (guess what) is fundamental to large parts of mathematics and even more so to physics. The politician's misconception was quickly pointed out … and he carried on *exactly as before.*

21 For technical reasons my remark about jigsaws doesn't solve the prize problem. If it did, I'd have got there first.

22 M.R. Garey and D.S. Johnson. *Computers and Intractability: A Guide to the Theory of NP-Completeness*, Freeman, San Francisco (1979).

23 G. Peano. 'Sur une courbe qui remplit toute une aire plane', *Mathematische Annalen* 36 (1890) 157–160.

24 Some care needs to be taken because some real numbers don't have unique representations as decimals – for instance $0\cdot500000\ldots = 0\cdot499999\ldots$. But that's easy to sort out.

25 E. Netto. 'Beitrag zur Mannigfaltigkeitslehre', *Journal für die Reine und Angewandte Mathematik* 86 (1879) 263–268.

26 H. Sagan. 'Some reflections on the emergence of space-filling curves: the way it could have happened and should have happened, but did not happen', *Journal of the Franklin Institute* 328 (1991) 419–430. For an explanation, see: A. Jaffer. 'Peano space-filling curves', http://people.csail.mit.edu/jaffer/Geometry/PSFC

27 J. Lawder. 'The application of space-filling curves to the storage and retrieval of multi-dimensional data', PhD Thesis, Birkbeck College, London (1999).

28 J. Bartholdi. 'Some combinatorial applications of spacefilling curves', www2.isye.gatech.edu/~jjb/research/mow/mow.html

29 H. Hahn. 'Über die allgemeinste ebene Punktmenge, die stetiges Bild einer Strecke ist', *Jahresbericht der Deutschen Mathematiker-Vereinigung,* 23 (1914) 318–322. H. Hahn. 'Mengentheoretische Charakterisierung der stetigen Kurven', *Sitzungsberichte der Kaiserlichen Akademie der Wissenschaften, Wien* 123 (1914) 2433–2489. S. Mazurkiewicz. 'O aritmetzacji kontinuów', *Comptes Rendus de la Société Scientifique de Varsovie* 6 (1913) 305–311 and 941–945.

30 Published in 1998: S. Arora, M. Sudan, R. Motwani, C. Lund, and M. Szegedy. 'Proof verification and the hardness of approximation problems', *Journal of the Association for Computing Machinery* 45 (1998) 501–555.

31 L. Babai. 'Transparent proofs and limits to approximation', in: *First European Congress of Mathematics. Progress in Mathematics* 3 (eds. A. Joseph, F. Mignot, F. Murat, B. Prum, and R. Rentschler) 31–91, Birkhäuser, Basel (1994).

32 C. Szegedy, W. Zaremba, I. Sutskever, J. Bruna, D. Erhan, I. Goodfellow, and R. Fergus. 'Intriguing properties of neural networks', arXiv:1312.6199 (2013).

33 A. Shamir, I. Safran, E. Ronen, and O. Dunkelman. 'A simple explanation for the existence of adversarial examples with small Hamming distance', arXiv:1901.10861v1 [cs.LG] (2019).

34 Not to be confused with the graph of a function, which is a curve relating a variable x to the value $f(x)$ of the function. Like the parabola for $f(x) = x^2$.

35 Thanks to Robin Wilson for gently pointing this out when I got it wrong in one of my books.

36 Provided you know which region to start from, it's enough just to list the bridge symbols, in the order they're crossed. Consecutive bridges determine a common region, to which they both connect.

37 This is fairly easy to prove using Euler's characterisation of open tours. The main idea is to break a hypothetical closed tour by cutting out one bridge. Now you have an open tour, and the bridge concerned originally joined the two ends.

38 The rest of this chapter is based on: D. Manlove. 'Algorithms for kidney donation', *London Mathematical Society Newsletter* 475 (March 2018) 19–24.

39 The exact date when Fermat stated his Last Theorem isn't certain, but it's often taken to be 1637.

40 The same can be said of much 'applied' mathematics too. However, there's a difference: the attitude of the mathematician. Pure mathematics is driven by the internal logic of the subject: not merely monkey curiosity, but a feeling for structure and a sense of where our understanding has significant gaps. Applied mathematics is mainly driven by problems arising in the 'real world', but it's more willing to tolerate unjustified shortcuts and approximations in search of an answer, and the answer may or may not have practical implications. As this chapter illustrates, however, a topic that seems completely useless at some moment in history can suddenly become vital to practical issues when culture or technology changes. Moreover, mathematics is an interconnected whole; even the pure/applied distinction is an artificial one. A theorem that seems useless in its own right may inspire, or even imply, results of great utility.

41 The answer is:
$p = 12,277,385,900,723,407,383,112,254,544,721,901,362,713,421,$
$995,519$
$q = 97,117,113,276,287,886,345,399,101,127,363,740,261,423,928,$
$273,451$
I found these two primes by trial and error, and multiplied them together, using a symbolic algebra system on a computer. This took a few minutes, mostly me changing digits at random until I stumbled across a prime. Then I told the computer to find the factors of the product, and it ran for ages with no result.

42 If n is a prime power p^k, then $\varphi(n) = p^k - p^{k-1}$. For a product of prime powers, multiply these expressions together for all the different prime powers in the prime factorisation of n. For instance, to find $\varphi(675)$ write $675 = 3^3 5^2$. Then
$$\varphi(675) = (3^3 - 3^2)(5^2 - 5) = (18)(20) = 360.$$

43 For more detail about the issues involved, see Ian Stewart, *Do Dice Play God?*, Profile, London (2019), Chapters 15 and 16.

44 L.M.K. Vandersypen, M. Steffen, G. Breyta, C.S. Yannoni, M.H. Sherwood, and I.L. Chuang. 'Experimental realization of Shor's quantum factoring algorithm using nuclear magnetic resonance', *Nature* 414 (2001) 883–887.

45 F. Arute and others. 'Quantum supremacy using a programmable superconducting processor', *Nature* 574 (2019) 505–510.

46 J. Proos and C. Zalka. 'Shor's discrete logarithm quantum algorithm for elliptic curves', *Quantum Information and Computation* 3 (2003).

47 M. Roetteler, M. Naehrig, K. Svore, and K. Lauter. 'Quantum resource estimates for computing elliptic curve discrete logarithms', in: *ASIACRYPT 2017: Advances in Cryptology*, Springer, New York (2017), 214–270.

48 For instance, -25 has a square root $5i$, because
$$(5i)^2 = 5i.5i = 5.5.i.i = 25i^2 = 25(-1) = -25$$
In fact, it has a second square root, $-5i$, for similar reasons.

49 Algebraists regularise the situation by saying that the square root of zero is zero, with *multiplicity* two. That is, the same value occurs twice, in a meaningful but technical sense. An expression like $x^2 - 4$ has two factors, $x + 2$ times $x - 2$, which respectively give two solutions $x = -2$ and $x = +2$ to the equation $x^2 - 4 = 0$. Similarly, the expression x^2 has two factors, x times x. They just happen to be the same.

50 For real c the function $z(t) = e^{ct}$ obeys the differential equation $dz/dt = cz$, with initial condition $z(0) = 1$. If we define the exponential function for complex c so that the same equation holds, which is sensible, and set $c = i$, then $dz/dt = iz$. Since multiplying by i rotates complex numbers through a right angle, the tangent to $z(t)$ as t varies is at right angles to $z(t)$, so the point $z(t)$ describes a circle of radius 1 centred at the origin. It rotates round this circle at a constant speed of one radian per unit of time, so at time t its position is at angle t radians. By trigonometry, this point is $\cos t + i \sin t$.

51 More precisely, there has to be an 'inner product', which determines distances and angles.

52 The fastest supercomputer in 1988 was the Cray Y-MP, costing $20 million (over $50 million in today's money). It would struggle to run a Windows operating system.

53 K. Shoemake. 'Animating rotation with quaternion curves', *Computer Graphics* 19 (1985) 245–254.

54 L. Euler. 'Découverte d'un nouveau principe de mécanique' (1752), *Opera Omnia, Series Secunda 5*, Orel Fusili Turici, Lausanne (1957), 81–108.

55 The half-angle property is important in quantum mechanics, where one formulation of quantum spin is based on quaternions. If the wave function of a particle of the kind known as a fermion is rotated through 360°, its spin reverses. (This is distinct from rotating the particle itself.) The wave function must rotate through 720° to return the spin to its original value. The unit quaternions form a 'double cover' of the rotations.

56 C. Brandt, C. von Tycowicz, and K. Hildebrandt. 'Geometric flows of curves in shape space for processing motion of deformable objects', *Computer Graphics Forum* 35 (2016) 295–305.

57 www.syfy.com/syfywire/it-took-more-cgi-than-you-think-to-bring-carrie-fisher-into-the-rise-of-skywalker

58 T. Takagi and M. Sugeno. 'Fuzzy identification of systems and its application to modeling and control', *IEEE Transactions on Systems, Man, and Cybernetics* 15 (1985) 116–132.

59 This is JFIF encoding, used for the web. Exif coding, for cameras, also includes 'metadata' describing the camera settings, such as date, time, and exposure.

60 A. Jain and S. Pankanti. 'Automated fingerprint identification and imaging systems', in: *Advances in Fingerprint Technology* (eds. C. Lee and R.E. Gaensslen), CRC Press, (2001) 275–326.

61 N. Ashby. 'Relativity in the Global Positioning System', *Living Reviews in Relativity* 6 (2003) 1; doi: 10.12942/lrr-2003-1.

62 More precisely, $Z=\sum\exp(-\beta H)$, where the sum is over all configurations of spin variables.

63 Setting $\beta = 1/k_B T$, where k_B is Boltzmann's constant, the formula is:

$$g(T, H) = -\frac{1}{\beta}\log\left[e^{\beta J}\cosh(\beta H) + \sqrt{e^{2\beta J}\cosh^2(\beta H) - 2\sinh(2\beta J)}\,\right]$$

64 The formula is:

$$\frac{\sinh(\beta H)}{\sqrt{\sinh^2(\beta H) + \exp(-4\beta J)}}$$

where H is the strength of the external field and J is the strength of the interactions between spins. In the absence of an external field $H = 0$, so $\sinh(\beta H) = 0$, so the whole fraction is 0.

65 Y.-P. Ma, I. Sudakov, C. Strong, and K.M. Golden. 'Ising model for melt ponds on Arctic sea ice', *New Journal of Physics* 21 (2019) 063029.

66 S. Tanaka. 'Topological analysis of point singularities in stimulus preference maps of the primary visual cortex', *Proceedings of the Royal Society of London B* 261 (1995) 81–88.

67 'Lobster telescope has an eye for X-rays', https://www.sciencedaily.com/releases/2006/04/060404194138.htm

68 Technically, the curve is the *image*, under a map from a disc to the sphere, of the boundary of the disc. The curve can cross itself and the disc can get scrunched up.

69 J.J. Berwald, M. Gidea, and M. Vejdemo-Johansson. 'Automatic recognition and tagging of topologically different regimes in dynamical systems', *Discontinuity, Nonlinearity, and Complexity* 3 (2014) 413–426.

70 F.A. Khasawneh and E. Munch. 'Chatter detection in turning using persistent homology', *Mechanical Systems and Signal Processing* 70 (2016) 527–541.

71 C.J. Tralie and J.A. Perea. '(Quasi) periodicity quantification in video data, using topology', *SIAM Journal on Imaging Science* 11 (2018) 1049–1077.

72 S. Emrani, T. Gentimis, and H. Krim. 'Persistent homology of delay embeddings and its application to wheeze detection', *IEEE Signal Processing Letters* 21 (2014) 459–463.

Picture Credits

Page 89: Tommy Muggleton (Redrawn).

Page 212: Jen Beatty. 'The Radon Transform and the Mathematics of Medical Imaging' (2012). *Honors Theses*. Paper 646. https://digitalcommons.colby.edu/honorstheses/646.

Page 238: Wikipedia.

Page 265: Yi-Ping Ma.

Page 276: G.G. Blasdel. 'Orientation selectivity, preference, and continuity in monkey striate cortex'. *Journal of Neuroscience* 12 (1992) 3139–3161.

Index

The index covers the man text but not the endnotes. Italic page references indicate relevant illustrations on pages which may not have relevant indexed text.

Index

Index